W9-AFU-623

ANTI-SCIENCE AND THE ASSAULT ON DEMOCRACY

ANTI-SCIENCE AND THE ASSAULT ON DEMOCRACY
DEFENDING REASON IN A FREE SOCIETY

EDITED BY

MICHAEL J. THOMPSON AND
GREGORY R. SMULEWICZ-ZUCKER

 Prometheus Books

59 John Glenn Drive
Amherst, New York 14228

Published 2018 by Prometheus Books

Anti-Science and the Assault on Democracy: Defending Reason in a Free Society. Copyright © 2018 by Michael J. Thompson and Gregory R. Smulewicz-Zucker. All rights reserved. No part of this publication may be reproduced, stored in a retrieval system, or transmitted in any form or by any means, digital, electronic, mechanical, photocopying, recording, or otherwise, or conveyed via the internet or a website without prior written permission of the publisher, except in the case of brief quotations embodied in critical articles and reviews.

Cover design by Jacqueline Nasso Cooke
Cover design © Prometheus Books

The internet addresses listed in the text were accurate at the time of publication. The inclusion of a website does not indicate an endorsement by the authors or by Prometheus Books, and Prometheus Books does not guarantee the accuracy of the information presented at these sites.

Inquiries should be addressed to
Prometheus Books
59 John Glenn Drive
Amherst, New York 14228
VOICE: 716–691–0133 • FAX: 716–691–0137
WWW.PROMETHEUSBOOKS.COM

22 21 20 19 18   5 4 3 2 1

Library of Congress Cataloging-in-Publication Data

Names: Thompson, Michael, 1973- editor. | Smulewicz-Zucker, Gregory R., 1983- editor.
Title: Anti-science and the assault on democracy : defending reason in a free society / edited by Michael J. Thompson and Gregory Smulewicz-Zucker.
Description: Amherst, New York : Prometheus Books, 2018. | "Defending the role that science must play in democratic society—science defined not just in terms of technology but as a way of approaching problems and viewing the world."
Identifiers: LCCN 2018021012 (print) | LCCN 2018030171 (ebook) |
    ISBN 9781633884755 (ebook) | ISBN 9781633884748 (hardcover)
Subjects: LCSH: Science—Political aspects. | Science—Social aspects.
Classification: LCC Q175.5 (ebook) | LCC Q175.5 .A58 2018 (print) | DDC 306.4/5—dc23
LC record available at https://lccn.loc.gov/2018021012

Printed in the United States of America

CONTENTS

PART IV: THE REVENGE OF ANTI-SCIENCE

INTRODUCTION

Acontradiction resides deep in the heart of modern society. On the one hand, we inhabit a world increasingly dominated by science and technology, one where the progress of scientific knowledge and technical efficiency seems without end. On the other hand, there also exists a deep-seated opposition to scientific knowledge and to science itself as a form of knowledge. A trend has been gathering momentum in modern culture away from science as a means to think about human affairs and an approach to truth. Although technology and technological forms of rationality have transformed our world, hostility toward science as a method and as a way of comprehending the social and natural world has emerged as an obstacle to a more humane and more democratic society. From religiously motivated arguments against the teaching of evolution in public schools to the denial of climate change, new-ageist espousals of alternative medicine, the regular distortion or dismissal of social-scientific data, outlandish claims about the effects of vaccinations or the fluoridation of water, and widespread basic ignorance about concepts such as "theory" or "evidence," anti-science viewpoints are becoming more and more manifest in our daily lives.

This trend is one that we call here "anti-science," and it is characterized by more than a skepticism of science as a body of knowledge about the natural world; it is also a hostility toward the very notion that objective truth claims can be defended. The anti-science attitude predisposes one to view science—as a mode of inquiry—as belonging solely to educated elites, who use it to "disenchant" the world and to control those with differing worldviews, particularly those, for example, who find their identity in knowledge that comes through nonscientific means. This anti-science position has its roots not only in the populist anti-elitism of the current period, it has also expressed itself in the halls of academe. The insidious influence of postmodern relativism equates the methods of the natural and social sciences with regimes of "power-knowledge" where all claims to rational, objective truth are questioned as mere masks for social power and dominance. A hyper-cynicism with respect to the powers of science, progress, and Enlightenment reason plagues many of those who consider themselves critics of modern society and politics. This trend has led to a convergence of traditionally conservative anti-science positions and Far-Left ideas about the social construction of knowledge.

In some ways this should be of little surprise. On the one hand, consumer society has increasingly isolated individuals from the implications and consequences of their preferences and actions. Each now sees their own subjective world as forming the common-sense context for their ideas about what matters and what does not. The internet and niche media outlets that foster the proverbial "echo chamber" have only exacerbated this social malaise. Isolated individuals enjoy the experience of community by finding online communities that reinforce their most outlandish beliefs and nurture even more bizarre worldviews. In addition, the proliferation of the self-help industry has combined the propensity for seeking irrationally grounded solutions with the tendency to portray all problems through the lens of a hyper-individualism, further encouraging people to withdraw into the self and away from the democratic community.

On the other hand, as modern societies become increasingly complex and require more technocratic management, there is an erosion of civic and democratic practices and mind-sets as people's relations become more mediated by technology and less by interpersonal contact. The result here is that technological society becomes bereft of value judgments as individuals rely less on their own ideas about what is right and wrong and more on the systems that shape their lives. As a result, society becomes impervious to the decisions and issues of citizens. Science as a form of objective reasoning about human affairs and the dynamics of the natural world now comes to be seen as tool of the elite that seeks to dislodge the individual from their place in their respective traditions and beliefs. It seems that, more and more, science has been misused by those in power, who seek to orient it toward profit, military power, or the efficiency of social control, just as democracy has withered—that science has been absorbed into technical rationality and forms of control rather than flourishing as a form of experiment and inquiry. Bringing these two streams of science and democracy together again is therefore no easy task, but it is essential to restate the basic linkage between science as a form of inquiry and democracy as a form of shared power for the common good of its members.

The relation between science and democracy has been evident since the origins of the modern world. The break with the medieval world order began to accelerate once skepticism toward traditional authority began to embed itself in the western mind. This skepticism had, at various historical moments, asserted itself in many parts of the non-Western world, such as within the Islamic world, but were tragically silenced by both religious and political authorities. By questioning both biblical ideas about nature and natural laws, as well as the encrusted hierarchical ideas of Christian philosophy or Scholasticism, science as an approach to knowledge stressed an intrinsic skepticism of any received truth and a rejection of any truth claim that was based on tradition or some social authority. What began in the Renaissance and continued on through the Enlightenment and the nineteenth century was a wedding between science and democratic attitudes. The progress of scientific reason was accompanied by movements toward representative democracy, the rule of law, and republicanism. But there was also a more subtle linkage formed between the attitudes of science and the attitudes of democratic life. Although the political and the scientific paths were distinct, each was demonstrating a new way to think about truth, about authority, and about the power of reason. What the scientific revolution and the age of democratic revolutions shared was a confidence in the human capacity to know, to create, and to test and to experiment with new ideas. As the sphere of the sacred began to shrink, the realm of the secular started to expand. With this shift there was a new sense of freedom—traditional constraints were cast off and a new vision of an emancipated future could be glimpsed.

Perhaps one of the most salient thinkers to wed the scientific and democratic positions was Benedict Spinoza. For Spinoza, the critique of nonrational and irrational beliefs was central to the cultivation of the free individual as well as a democratic community. He advocated the thesis that the religious conception of God was a mythical misunderstanding of nature and that science therefore has the possibility to free us from superstition and provide us with the capacity to make self-governing decisions about our lives, both personally and collectively. Spinozist ideas spread throughout Europe and influenced many of the radical and revolutionary ideas of the eighteenth century. They nourished an array of brilliant thinkers for almost two centuries. But soon the backlash was to begin. Movements against Enlightenment reason from forces such as the church and feudal powers began to weave a new narrative about the destructive tendencies of science and democracy. The reactionary Counter-Enlightenment—made up of figures throughout Europe such as Joseph de Maistre, Edmund Burke, and J. G. Hamann—argued that the dual influences of modern science and democratic government were threats to the vital traditions and forms of life that gave society its sense of structure and stability. Tradition and hierarchy, throne and altar were now to be counterposed against the emerging forces of science, reason, and democratic life.

The forces of reaction have been tied with the forces of Enlightenment in a kind of Jacob-and-Angel struggle ever since. But recent years have displayed a new kind of anti-scientific sensibility. As modern market societies dominated by capitalism have fragmented social bonds and eroded social trust, and the infantilization of mass-produced popular culture has been successful in derationalizing citizens, irrational forms of meaning are sought for. Mysticism, a return to religion, nihilism, the preponderance of self-help manuals, new-age gurus—all are dimensions of modern culture that threaten to render democratic life sterile and inert in the face of new forms of inequality and authority. Any hope for a revival of democratic society must therefore be premised on a rebirth of a kind of scientific attitude. One reason for this is that, with the collapse of organized religious doctrines and broadly shared traditions and values, modern society can be democratically sewn together via forms of shared norms and institutions that root their legitimacy in our rational and reflective capacity for consent. Science as a mode of inquiry is uniquely suited to this task. In its emphasis on reason, evidence, and revisability, it can mold the mind into a form appropriate for this kind of democratic life.

The problem is, however, that the concept of democracy that was informed by Enlightenment ideas has been on the decline in modern societies. Democracy is now becoming viewed not as a process of determining and consenting to common, objective solutions to social problems and needs, but as a domain to express one's emotions and where each member's belief systems are to have as much warrant as any other. Instead of breaking down parochial belief systems, we seem to be cast into a Babel-like clash of worldviews. It is our contention that tendencies that defend anti-science views in the name of "democratic pluralism"—whether invoked by the Left or the Right—are actually destructive of a broader culture and the mind-set appropriate for a democratic society. Rather than enhancing the capacity for rational debate and critical discourse, we could see this as a return to premodern forms of subservience to authority and a deeply entrenched irrational refusal to submit beliefs to rational scrutiny. In this respect, we believe that we are witnessing a peculiarly new manifestation of anti-science belief systems that are, in spirit, akin to a premodern worldview but carry new consequences in contemporary society and politics.

The debate that the essays in this volume address is, in a certain respect, an old one. Among its most recent manifestations were the so-called "science wars" of the 1990s, which pitted defenders of scientific rationality against postmodernists and relativists. This was coupled with the efforts on the part of conservative evangelical Christians, whose new political self-assertiveness was a product of Ronald Reagan's election, to remount the battle against the teaching of evolution. The book therefore picks up on these older debates insofar as a certain brand of postmodern thinking remains prevalent in academia and the pseudo debate over the teaching of evolution persists and is now joined by climate-change denial. However, we believe that current manifestations of anti-science views have helped to forge a political crisis. Not only have the anti-science views of postmodernists come to inform democratic political theory, but also these more academic concerns complement a broader political climate in which a resistance to science shapes public opinion and a more general hostility to scientific reason.

The defense of creationism and resistance to the realities of climate change, for example, threaten both the capacity of the next generation to comprehend and employ rational arguments as well as its ability to confront the ways anti-science worldviews prevent us from dealing with environmental crises. In effect, postmodern theories on the Left and religiously influenced policy proposals on the Right have become unlikely allies with one another, even if not in intent. For these reasons, the essays in this volume put a particular emphasis on the political implications of anti-science thought, while the older "science wars" were conducted around more philosophical questions of method and epistemology. The new realities of anti-science policy—the withdrawal of funding from agencies that rely on the sciences, the effort to redesign school curricula, and many related concerns—speak to a political terrain in which anti-science views are reshaping our institutions, our culture, and our sensibilities.

Moreover, the issue of an anti-scientific worldview and its accompanying dismissal of the importance of reason have had an impact on contemporary political discourse. In a context where notions such as "alternative facts" have helped to foster populist political sentiment, we believe that the defense of a scientific worldview helps to oppose this broader social tendency to privilege ungrounded belief over reasoned argument and evidence. As we are now witnessing, this carries dangerous consequences for environmental policy and the health and well-being of citizens. Such political trends, we argue, are symptomatic of the bolstering of modes of thought that reject the values behind good science. Hence, when contributors to this volume speak of science, they not only have in mind knowledge of the laws and theories in the natural sciences, but, in addition, the type of intellectual procedures and forms of inquiry upon which the discovery of such laws and the development of such theories are founded. Viewed from this vantage point, postmodernist claims about the overlapping of power and knowledge finds their political expression in a more general tendency to view expertise as a form of elitism and the arguments of scientists as an attempt to forcibly impose beliefs. The degradation of science in academia, at best, leaves the next generation ignorant and, at worst, incapacitates the ability to combat irrationalism in civil society.

This is a problem that has only gotten worse with time. As democratic engagement and attitudes have eroded in American society so has investment in public education. Our public school systems are being starved of funding, and students are forced to share outdated textbooks that are literally coming apart at the seams. Meanwhile, wealthy donors with fundamentalist beliefs—whether of religion or the market—are seeking to lure the most vulnerable students into charter schools where they can more easily influence the curriculum. At the same time, they seek to use their wealth to infiltrate the boards of our great public institutions that sought to inspire wonder and curiosity in all citizens, regardless of class, such as the American Museum of Natural History. Add to this the spreading phenomenon of the corrosive effects of a commodified popular culture and we begin to grasp the extent to which reason—the cradle of science—has been receding. This culture of commodification has even transformed the popular mind's conception of what constitutes intellectual achievement, with more and more young people looking to Silicon Valley elites and the production of ever-more time-wasting features for their smartphones as marks of intellectual contributions to the social good.

Technological society has become increasingly dominant and with it an atrophy of the individual's moral and cognitive faculties. Students at universities can scarcely endure their lectures without attempting to clandestinely play the latest video game on their phone. Adults, in turn, are infantilized by these same technologies and are as addicted to social media as young people. The dominance of technique—with its emphasis on means over ends, prescribed processes over creative experiment, and conformity over creativity and spontaneity—means that the well-springs of political judgment are drying up. With these new realities in mind, the basic question and problem the essays assembled here seek to examine is the extent to which we can draw upon the relationship between science and democracy. More specifically, we want to explore the ways that science—as an attitude as well as an epistemic style of thought—is in retreat in modern societies. We therefore seek not only to polemically critique the anti-scientific cast of different features of modern culture and politics, but also to stake out a defense of rational, scientific dimensions to democratic life and a democratic society.

The essays collected here probe the problem of how these trends are affecting democratic society. The collective thesis of the essays that follow maintains that democracy is withering due in no small part to our culture's drifting away from the scientific mind-set and the erosion of reason that results from it. What is the relation between science and democracy? What is the difference between science and its technological-industrial expressions? What are the consequences and implications of opposition to science for political life? How does a rational-scientific attitude aid in core democratic values such as objective social knowledge, tolerance, just institutions, the promotion of a common interest, and so on? In the end, the fundamental question is: How does science—and the distinctive form of rationality that it promotes—underwrite a democratic society? For the essays in this book, science is more than a technical enterprise, it is an attitude toward reason, toward the importance of objective knowledge and the kind of inquiry that calls into question the traditions and practices taken for granted or as without justification by the community as a whole or by one's tradition or group in particular. We therefore insist that science, in the end, must be viewed as a core feature of a modern, democratic society and that its erosion in our culture is having, and will continue to have, grave political consequences. If the essays presented here go in any small way toward a reconstruction of democracy, we will all be the better for it.

PART I

REFORGING THE LINK BETWEEN
SCIENCE AND DEMOCRACY

CHAPTER 1

WHAT IS SCIENCE AND WHY SHOULD WE CARE?

ALAN SOKAL

I propose to share with you a few reflections about the nature of scientific inquiry and its importance for public life. At a superficial level one could say that I will be addressing some aspects of the relation between science and society; but, as I hope will become clear, my aim is to discuss the importance, not so much of science, but of what one might call the scientific worldview[1]—a concept that goes far beyond the specific disciplines that we usually think of as "science"—in humanity's collective decision-making. I want to argue that clear thinking, combined with a respect for evidence—especially inconvenient and unwanted evidence, evidence that challenges our preconceptions—are of the utmost importance to the survival of the human race in the twenty-first century, and especially so in any polity that professes to be a democracy.

Of course, you might think that calling for clear thinking and a respect for evidence is a bit like advocating Motherhood and Apple Pie (if you'll pardon me this Americanism)—and in a sense you'd be right. Hardly anyone will openly defend muddled thinking or disrespect for evidence. Rather, what people do is to surround these confused practices with a fog of verbiage designed to conceal from their listeners—and in most cases, I would imagine, from themselves as well—the true implications of their way of thinking. George Orwell got it right when he observed that the main advantage of speaking and writing clearly is that "when you make a stupid remark its stupidity will be obvious, even to yourself."[2] So I hope that I will be as clear tonight as Orwell would have wished. And I intend to illustrate disrespect for evidence with a variety of examples—coming from the Left and the Right and the Center—starting from some fairly lightweight targets and proceeding to heavier ones. I aim to show that the implications of taking seriously an evidence-based worldview are rather more radical than many people realize.

So let me start by drawing some important distinctions. The word science, as commonly used, has at least four distinct meanings: it denotes an intellectual endeavor aimed at a rational understanding of the natural and social world; it denotes a corpus of currently accepted substantive knowledge; it denotes the community of scientists, with its mores and its social and economic structure; and, finally, it denotes applied science and technology. In this essay I will be concentrating on the first two aspects, with some secondary references to the sociology of the scientific community; I will not address technology at all. Thus, by science I mean, first of all, a worldview giving primacy to reason and observation and, second, a methodology aimed at acquiring accurate knowledge of the natural and social world. This methodology is characterized, above all else, by the critical spirit: namely, the commitment to the incessant testing of assertions through observations and/or experiments—the more stringent the tests, the better—and to revising or discarding those theories that fail the test.[3] One corollary of the critical spirit is fallibilism: namely, the understanding that all our empirical knowledge is tentative, incomplete, and open to revision in the light of new evidence or cogent new arguments (though, of course, the most well-established aspects of scientific knowledge are unlikely to be discarded entirely).

It is important to note that well-tested theories in the mature sciences are supported in general by a powerful web of interlocking evidence coming from a variety of sources. Moreover, the progress of science tends to link these theories into a unified framework, so that (for instance) biology has to be compatible with chemistry, and chemistry with physics. The philosopher Susan Haack has illuminatingly analogized science to the problem of completing a crossword puzzle, in which any modification of one word will entail changes in interlocking words; in most cases the required changes will be fairly local, but in some cases it may be necessary to rework large parts of the puzzle.[4]

I stress that my use of the term "science" is not limited to the natural sciences, but includes investigations aimed at acquiring accurate knowledge of factual matters relating to any aspect of the world by using rational empirical methods analogous to those employed in the natural sciences. (Please note the limitation to questions of fact. I intentionally exclude from my purview questions of ethics, aesthetics, ultimate purpose, and so forth.) Thus, "science" (as I use the term[5]) is routinely practiced not only by physicists, chemists, and biologists, but also by historians, detectives, plumbers, and indeed all human beings in (some aspects of) our daily lives.[6] (Of course, the fact that we all practice science from time to time does not mean that we all practice it equally well, or that we practice it equally well in all areas of our lives.)

The extraordinary successes of the natural sciences over the last four hundred years in learning about the world, from quarks to quasars and everything in between, are well known to every modern citizen: science is a fallible yet enormously successful method for obtaining objective (albeit approximate and incomplete) knowledge of the natural (and to a lesser extent, the social) world.

But, surprisingly, not everyone accepts this; and here I come to my first—and most lightweight—example of adversaries of the scientific worldview, namely academic postmodernists and extreme social constructivists. Such people insist that so-called scientific knowledge does not in fact constitute objective knowledge of a reality external to ourselves, but is a mere social construction, on a par with myths and religions, which therefore have an equal claim to validity. If such a view seems so implausible that you wonder whether I am somehow exaggerating, consider the following assertions by prominent sociologists:

> "The validity of theoretical propositions in the sciences is in no way affected by factual evidence." (Kenneth Gergen)[7]
> "The natural world has a small or non-existent role in the construction of scientific knowledge." (Harry Collins)[8]
> "For the relativist [such as ourselves] there is no sense attached to the idea that some standards or beliefs are really rational as distinct

from merely locally accepted as such." (Barry Barnes and David Bloor)[9]

"Since the settlement of a controversy is the cause of Nature's representation not the consequence, we can never use the outcome— Nature—to explain how and why a controversy has been settled." (Bruno Latour)[10]

"Science legitimates itself by linking its discoveries with power, a connection which determines (not merely influences) what counts as reliable knowledge." (Stanley Aronowitz)[11]

Statements as clear-cut as these are, however, rare in the academic postmodernist literature. More often one finds assertions that are ambiguous but can nevertheless be interpreted (and quite often are interpreted) as implying what the foregoing quotations make explicit: that science as I have defined it is an illusion, and that the purported objective knowledge provided by science is largely or entirely a social construction. For example, Katherine Hayles, professor of literature at Duke University and former president of the Society for Literature and Science, writes the following as part of her feminist analysis of fluid mechanics:

Despite their names, conservation laws are not inevitable facts of nature but constructions that foreground some experiences and marginalize others. . . . Almost without exception, conservation laws were formulated, developed, and experimentally tested by men. If conservation laws represent particular emphases and not inevitable facts, then people living in different kinds of bodies and identifying with different gender constructions might well have arrived at different models for [fluid] flow.[12]

(What an interesting idea: perhaps "people living in different kinds of bodies" will learn to see beyond those masculinist laws of conservation of energy and momentum.) And Andrew Pickering, a prominent sociologist of science, asserts the following in his otherwise-excellent history of modern elementary-particle physics:

Given their extensive training in sophisticated mathematical techniques, the preponderance of mathematics in particle physicists'
accounts of reality is no more hard to explain than the fondness of
ethnic groups for their native language. On the view advocated in this
chapter, there is no obligation upon anyone framing a view of the world
to take account of what twentieth-century science has to say.[13]

But let me not spend time beating a dead horse, as the arguments
against postmodernist relativism are by now fairly well known: rather than
plugging my own writings, let me suggest the superb book by the Canadian philosopher of science James Robert Brown, *Who Rules in Science? An
Opinionated Guide to the Wars*.[14] Suffice it to say that postmodernist writings systematically confuse truth with claims of truth, fact with assertions
of fact, and knowledge with pretensions to knowledge—and then sometimes go so far as to deny that these distinctions have any meaning.

Now, it's worth noting that the postmodernist writings I have just
quoted all come from the 1980s and early 1990s. In fact, over the past
decade, academic postmodernists and social constructivists seem to have
backed off from the most extreme views that they previously espoused.
Perhaps I and like-minded critics of postmodernism can take some small
credit for this, through initiating a public debate that shed a harsh light
of criticism on these views and forced some strategic retreats. But most of
the credit, I think, has to be awarded to George W. Bush and his friends,
who showed just where science bashing can lead in the real world.[15]
Nowadays, even sociologist of science Bruno Latour, who spent several
decades stressing the so-called "social construction of scientific facts,"[16]
laments the ammunition he fears he and his colleagues have given to the
Republican right wing, helping them to deny or obscure the scientific
consensus on global climate change, biological evolution, and a host of
other issues.[17] He writes:

> While we spent years trying to detect the real prejudices hidden behind
> the appearance of objective statements, do we now have to reveal the
> real objective and incontrovertible facts hidden behind the illusion of
> prejudices? And yet entire PhD programs are still running to make sure

that good American kids are learning the hard way that facts are made up, that there is no such thing as natural, unmediated, unbiased access to truth, that we are always prisoners of language, that we always speak from a particular standpoint, and so on, while dangerous extremists are using the very same argument of social construction to destroy hard-won evidence that could save our lives.[18]

That, of course, is exactly the point I was trying to make back in 1996 about social construction talk taken to subjectivist extremes. I hate to say I told you so, but I did—as did, several years before me, Noam Chomsky, who recalled that in a not-so-distant past, left-wing intellectuals took an active part in the lively working-class culture. Some sought to compensate for the class character of the cultural institutions through programs of workers' education, or by writing bestselling books on mathematics, science, and other topics for the general public. Remarkably, their left-wing counterparts today often seek to deprive working people of these tools of emancipation, informing us that the "project of the Enlightenment" is dead, that we must abandon the "illusions" of science and rationality—a message that will gladden the hearts of the powerful, who are delighted to monopolize these instruments for their own use.[19]

Let me now pass to a second set of adversaries of the scientific worldview, namely the advocates of pseudoscience.[20] This is of course an enormous area, so let me focus on one socially important aspect of it, namely so-called "complementary and alternative therapies" in health and medicine. And within this, I'd like to look in a bit of detail at one of the most widely used "alternative" therapies, namely homeopathy—which is an interesting case because its advocates sometimes claim that there is evidence from meta-analyses of clinical trials that homeopathy works.

Now, one basic principle in all of science is GIGO: garbage in, garbage out. This principle is particularly important in statistical meta-analysis because if you have a bunch of methodologically poor studies, each with small sample size, and then subject them to meta-analysis, what can happen is that the systematic biases in each study—if they mostly point in the same direction—can reach statistical significance when the studies are pooled. And this possibility is particularly relevant here because meta-analyses of homeopathy invariably find an inverse correlation between the methodological quality of the study and the observed effectiveness of homeopathy: that is, the sloppiest studies find the strongest evidence in favor of homeopathy.[21] When one restricts attention only to methodologically sound studies—those that include adequate randomization and double-blinding, predefined outcome measures, and clear accounting for dropouts—the meta-analyses find no statistically significant effect (whether positive or negative) of homeopathy compared to placebo.[22]

But the lack of convincing statistical evidence for the efficacy of homeopathy is not, in fact, the main reason why I and other scientists are skeptical (to put it mildly) about homeopathy; and it's worth taking a few moments to explain this main reason because it provides some important insights into the nature of science. Most people—perhaps even most users of homeopathic remedies—do not clearly understand what homeopathy is. They probably think of it as a species of herbal medicine. Of course plants contain a wide variety of substances, some of which can be biologically active (with either beneficial or harmful consequences, as Socrates learned). But homeopathic remedies, by contrast, are pure water and starch: the alleged "active ingredient" is so highly diluted that in most cases not a single molecule remains in the final product.

And so, the fundamental reason for rejecting homeopathy is that there is no plausible mechanism by which homeopathy could possibly work, unless one rejects everything that we have learned over the last two hundred years about physics and chemistry: namely, that matter is made of atoms, and that the properties of matter—including its chemical and biological effects—depend on its atomic structure. There is simply no way that an absent "ingredient" could have a therapeutic effect. High-quality clinical trials find no difference between homeopathy and placebo because homeopathic remedies are placebos.[23]

Now, advocates of homeopathy sometimes respond to this argument by asserting that the curative effect of homeopathic remedies arises from a "memory" of the vanished active ingredient that is somehow retained by the water in which it was dissolved (and then by the starch when the water is evaporated!). But the difficulty, once again, is not simply the lack of any reliable experimental evidence for such a "memory of water." Rather, the problem is that the existence of such a phenomenon would contradict well-tested science, in this case the statistical mechanics of fluids. The molecules of any liquid are constantly being bumped by other molecules—what physicists call thermal fluctuations—so that they quickly lose any "memory" of their past configuration. (Here when I say "quickly," I'm talking picoseconds, not months.)

In short, all the millions of experiments confirming modern physics and chemistry also constitute powerful evidence against homeopathy. For this reason, the flaw in the justification of homeopathy is not merely the lack of statistical evidence showing the efficacy of homeopathic remedies over placebo at the 95 percent or 99 percent confidence level. Even a clinical trial at the 99.99 percent confidence level would not begin to compete with all the evidence in favor of modern physics and chemistry. Extraordinary claims require extraordinary evidence. (And in the unlikely event that such convincing evidence is ever forthcoming, the person who provides it will assuredly win a triple Nobel Prize in physics, chemistry, and biology—beating out Marie Curie, who won only two.)

Despite the utter scientific implausibility of homeopathy, homeopathic products can be marketed in the United States without having to meet the safety and efficacy requirements that are demanded of all other drugs (because they got a special dispensation in the Food, Drug, and Cosmetic Act of 1938). Indeed, US government regulations require each homeopathic remedy that is marketed over-the-counter (OTC) to state, on the label, at least one medical condition that the product is intended to treat—but without requiring any evidence that the product is actually efficacious in treating that condition![24] The laws in other Western countries are equally scandalous, if not more so.[25]

Fortunately, it seems that this particular pseudoscience has thus far made only modest inroads in the United States—in contrast to its wide penetration in France and Germany, where homeopathic products are

packaged like real medicines and sold side by side with them in virtually every pharmacy. But other and more dangerous pseudosciences are
endemic in the United States: prominent among these is the denial of
biological evolution. It is essential to begin our analysis by distinguishing
clearly between three very different issues: namely, the fact of the evolution of biological species; the general mechanisms of that evolution; and
the precise details of those mechanisms. Of course, one of the favorite
tactics of deniers of evolution is to confuse these three aspects.

Among biologists, and indeed among the general educated public,
the fact that biological species have evolved is established beyond any reasonable doubt. Most species that existed at various times in the past no
longer exist; and conversely, most species that exist today did not exist
for most of the earth's past. In particular, modern *Homo sapiens* did not
exist one million years ago, and conversely, other species of hominids,
such as *Homo erectus*, existed then and are now extinct. The fossil record
is unequivocal on this point, and this has been well understood since at
least the late nineteenth century.

A more subtle issue concerns the mechanisms of biological evolution; and here our modern scientific understanding took a longer time to
develop. Though the basic idea—descent with modification, combined
with natural selection—was set forth with eminent clarity by Darwin in
his 1859 book, *On the Origin of Species*, the precise mechanisms underlying Darwinian evolution were not fully elucidated until the development of genetics and molecular biology in the first half of the twentieth
century. Nowadays we have a good understanding of the overall process:
errors in copying DNA during reproduction cause mutations; some of
these mutations either increase or decrease the organism's success at survival and reproduction; natural selection acts to increase the frequency in
the gene pool of those mutations that increase the organism's reproduc-
tive success; as a result, over time, species develop adaptations to ecological niches; old species die out and new species arise. This general picture
is nowadays established beyond any reasonable doubt, not only by pale-
ontology but also by laboratory experiments.

Of course, when it comes to the precise details of evolutionary theory, there is still lively debate among specialists (just as there is in any active scientific field): for instance, concerning the quantitative importance of group selection or of genetic drift. But these debates in no way cast doubt on either the fact of evolution or on its general mechanisms. Indeed, as the celebrated geneticist Theodosius Dobzhansky pointed out in a 1973 essay, "nothing in biology makes sense except in the light of evolution."[26]

Everything that I have just said is, of course, common knowledge to anyone who has taken a half-decent course in high-school biology. The trouble is, fewer and fewer people—at least in the United States—nowadays have the good fortune to be exposed to a half-decent course in high-school biology. And the cause of that scientific illiteracy is (need I say it?) politics: more precisely, politics combined with religion. Some people reject evolution because they find it incompatible with their religious beliefs. And in countries where such people are numerous or politically powerful or both, politicians kowtow to them and suppress the teaching of evolution in the public schools—with the result that the younger generation is denied the opportunity to evaluate the scientific evidence for themselves, and the scientific ignorance of the populace is faithfully[27] reproduced in future generations.

In 2005, a fascinating cross-cultural survey was carried out in thirty-two European countries, along with the United States and Japan.[28] Respondents were read the statement, "Human beings, as we know them, developed from earlier species of animals," and were asked whether they considered it to be true, false, or were not sure. Of all thirty-four countries, the United States holds thirty-third place for belief in evolution (with roughly equal numbers responding "true" and "false"). Only Turkey—where the secular heritage is under increasing assault from the elected Islamist government and its supporters—shows less belief in evolution than the United States. (Please note that this question concerns merely the fact of evolution, not its mechanisms.)

Of course, not all religious people reject evolution. Fundamentalist Christians do reject evolution, as do many Muslims and orthodox Jews, but Catholics and liberal Protestants have come (over time and perhaps grudgingly) to accept evolution, as have some Muslims and most Jews.[29] Therefore, from a purely tactical point of view, nonfundamentalist religious people are the allies of scientists in their struggle to defend the honest teaching of science.

And so, if I were tactically minded, I would stress—as most scientists do—that science and religion need not come into conflict. I might even go on to argue, following Stephen Jay Gould, that science and religion should be understood as "nonoverlapping magisteria": science dealing with questions of fact, religion dealing with questions of ethics and meaning.[30] But I can't in good conscience proceed in this way, for the simple reason that I don't think the arguments stand up to careful logical examination. Why do I say that? For the details, I have to refer you to a seventy-five-page chapter in my book,[31] but let me at least try to sketch now the main reasons why I think that science and religion are fundamentally incompatible ways of looking at the world.[32]

When analyzing religion, a few distinctions are perhaps in order. For starters, religious doctrines typically have two components: a factual part, consisting of a set of claims about the universe and its history; and an ethical part, consisting of a set of prescriptions about how to live. In addition, all religions make, at least implicitly, epistemological claims concerning the methods by which humans can obtain reasonably reliable knowledge of factual or ethical matters. These three aspects of each religion obviously need to be evaluated separately.

Furthermore, when discussing any set of ideas, it is important to distinguish between the intrinsic merit of those ideas, the objective role they play in the world, and the subjective reasons for which various people defend or attack them.

Alas, much discussion of religion fails to make these elementary distinctions: for instance, confusing the intrinsic merit of an idea with the good or bad effects that it may have in the world. Here I want to address only the most fundamental issue, namely, the intrinsic merit of the various religions' factual doctrines. And within that, I want to focus on the epistemological question—or to put it in less fancy language, the relationship between belief and evidence. After all, those who believe in their religion's factual doctrines presumably do so for what they consider to be good reasons. So it's sensible to ask: What are these alleged good reasons?

Each religion makes scores of purportedly factual assertions about everything from the creation of the universe to the afterlife. But on what grounds can believers presume to know that these assertions are true? The reasons they give are various, but the ultimate justification for most religious people's beliefs is a simple one: we believe what we believe because our holy scriptures say so. But how, then, do we know that our holy scriptures are factually accurate? Because the scriptures themselves say so.[33] Theologians specialize in weaving elaborate webs of verbiage to avoid saying anything quite so bluntly, but this gem of circular reasoning really is the epistemological bottom line on which all "faith" is grounded. In the words of Pope John Paul II, "By the authority of his absolute transcendence, God who makes himself known is also the source of the credibility of what he reveals."[34] It goes without saying that this begs the question of whether the texts at issue really were authored or inspired by God, and on what grounds one knows this. "Faith" is not in fact a rejection of reason but simply a lazy acceptance of bad reasons. "Faith" is the pseudo-justification that some people trot out when they want to make claims without the necessary evidence.

But of course we never apply these lax standards of evidence to the claims made in the other fellow's holy scriptures: when it comes to religions other than one's own, religious people are as rational as everyone else. Only our own religion, whatever it may be, seems to merit some special dispensation from the general standards of evidence. And here, it seems to me, is the crux of the conflict between religion and science. Not the religious rejection of specific scientific theories (be it heliocentrism in the seventeenth century or evolutionary biology today); over time most religions do find some way to make peace with well-established science. Rather, the scientific worldview and the religious worldview come into conflict over a far more fundamental question: namely, what constitutes evidence.

Science relies on publicly reproducible sense experience (that is, experiments and observations) combined with rational reflection on those empirical observations. Religious people acknowledge the validity of that method, but then claim to be in the possession of additional methods for obtaining reliable knowledge of factual matters—methods that go beyond the mere assessment of empirical evidence—such as intuition, revelation, or the reliance on sacred texts. But the trouble is this: What good reason do we have to believe that such methods work, in the sense of steering us systematically (even if not invariably) toward true beliefs rather than toward false ones?[35] At least in the domains where we have been able to test these methods—astronomy, geology, and history, for instance—they have not proven terribly reliable. Why should we expect them to work any better when we apply them to problems that are even more difficult, such as the fundamental nature of the universe?

Last but not least, these nonempirical methods suffer from an insurmountable logical problem: What should we do when different people's intuitions or revelations conflict? How can we know which of the many purportedly sacred texts—whose assertions frequently contradict one another—are in fact sacred?

In all these examples I have been at pains to distinguish clearly between factual matters and ethical or aesthetic matters, because the epistemological issues they raise are so different. And I have restricted my discussion almost entirely to factual matters, simply because of the limitations of my own competence. But if I am preoccupied by the relation between belief and evidence, it is not solely for intellectual reasons—not solely because I'm a "grumpy old fart who aspire[s] to the sullen joy of having it known that [I] don't suffer fools gladly"[36] (to borrow the words of my friend and fellow gadfly Norm Levitt, who died suddenly four years ago at the young age of sixty-six). Rather, my concern that public debate be grounded in the best available evidence is, above all else, ethical.

To illustrate the connection I have in mind between epistemology and ethics, let me start with a fanciful example: Suppose that the leader of a militarily powerful country believes, sincerely but erroneously, on the basis of flawed "intelligence," that a smaller country possesses threatening weapons of mass destruction; and suppose further that he launches a preemptive war on that basis, killing tens of thousands of innocent civilians as "collateral damage." Aren't he and his supporters ethically culpable for their epistemic sloppiness?

I stress that this example is fanciful. The overwhelming preponderance of currently available evidence suggests that the Bush and Blair administrations first decided to overthrow Saddam Hussein, and then sought a publicly presentable pretext, using dubious or even forged "intelligence" to "justify" that pretext and to mislead Congress, Parliament, and the public into supporting that war.[37]

Which brings me to the last, and in my opinion most dangerous, set of adversaries of the evidence-based worldview in the contemporary world: namely, propagandists, public-relations flacks, and spin doctors, along with the politicians and corporations who employ them—in short, all those whose goal is not to analyze honestly the evidence for and against a particular policy, but is, rather, simply to manipulate the public into reaching a predetermined conclusion by whatever technique will work, however dishonest or fraudulent.

So the issue here is no longer mere muddled thinking or sloppy reasoning; it is fraud.

The *Oxford English Dictionary* defines "fraud" as "the using of false representations to obtain an unjust advantage or to injure the rights or interests of another."[38] In the Anglo-American common law, a "false representation" can take many forms, including:

- A false statement of fact, known to be false at the time it was made;[39]
- A statement of fact with no reasonable basis to make that statement;[40]
- A promise of future performance made with an intent, at the time the promise was made, not to perform as promised;[41]
- An expression of opinion that is false, made by one claiming or implying to have special knowledge of the subject matter of the opinion—where "special knowledge" means knowledge or information superior to that possessed by the other party, and to which the other party did not have equal access.[42]

Anything here sound familiar? These are the standards that we would use if George Bush and Tony Blair had sold us a used car. In fact, they sold us a war that has cost the lives of 179 British soldiers, 4,486 American soldiers, and somewhere between 112,000 and 600,000 Iraqis[43]—a human toll, that is, of somewhere between 35 and 200 September 11ths; that has cost the American taxpayers a staggering $810 billion (with ultimate estimates ranging from $1–3 trillion);[44] and that has strengthened both al-Qaeda and Iran—in short, a war that may well turn out to be the greatest foreign-policy blunder of American history. (Of course the British have a longer history, and hence a longer history of blunders to compete with.)

Now, in the common law there are in fact two distinct torts of misrepresentation: negligent misrepresentation and fraudulent misrepresentation. Fraudulent misrepresentation is of course difficult to prove because it involves the state of mind of the person making the misrepresentation, i.e., what he actually knew or believed at the time of the false statement.[45] Which means that the question becomes (as it was in the case of an earlier American president who stood accused of far lesser crimes and misdemeanors): What did George Bush and Tony Blair know and when did they know it? Unfortunately, the documents that could elucidate this question are top secret, so we may not know the answer for fifty years, if ever. But enough documents have been leaked so far to support, I think, a verdict of fraudulent misrepresentation.[46]

Now, all this is very likely old hat to most readers. We know perfectly well that our politicians (or at least some of them) lie to us; we take it for granted; we are inured to it. And that may be precisely the problem. Perhaps we have become so inured to political lies—so hard-headedly cynical—that we have lost our ability to become appropriately outraged. We have lost our ability to call a spade a spade, a lie a lie, a fraud a fraud. Instead we call it "spin."[47]

We have now traveled a long way from "science," understood narrowly as physics, chemistry, biology, and the like. But the whole point is that any such narrow definition of science is misguided. We live in a single real world; the administrative divisions used for convenience in our universities do not in fact correspond to any natural philosophical boundaries. It makes no sense to use one set of standards of evidence in physics, chemistry, and biology, and then suddenly relax your standards when it comes to medicine, religion, or politics. Lest this sound to you like a scientist's imperialism, I want to stress that it is exactly the contrary. As the philosopher Susan Haack lucidly observes,

> Our standards of what constitutes good, honest, thorough inquiry and what constitutes good, strong, supportive evidence are not internal to science. In judging where science has succeeded and where it has failed, in what areas and at what times it has done better and in what worse, we are appealing to the standards by which we judge the solidity of empirical beliefs, or the rigor and thoroughness of empirical inquiry, generally.[48]

The bottom line is that science is not merely a bag of clever tricks that turn out to be useful in investigating some arcane questions about the inanimate and biological worlds. Rather, the natural sciences are nothing more or less than one particular application—albeit an unusually successful one—of a more general rationalist worldview, centered on the modest insistence that empirical claims must be substantiated by empirical evidence.

Conversely, the philosophical lessons learned from four centuries of work in the natural sciences can be of real value—if properly understood—in other domains of human life. Of course, I am not suggesting that historians or policymakers should use exactly the same methods as physicists—that would be absurd. But neither do biologists use precisely the same methods as physicists; nor, for that matter, do biochemists use the same methods as ecologists, or solid-state physicists as elementary-particle physicists. The detailed methods of inquiry must of course be adapted to the subject matter at hand. What remains unchanged in all areas of life, however, is the underlying philosophy: namely, to constrain our theories as strongly as possible by empirical evidence, and to modify or reject those theories that fail to conform to the evidence. That is what I mean by the scientific worldview.

It is because of this general philosophical lesson, far more than any specific discoveries, that the natural sciences have had such a profound effect on human culture since the time of Galileo and Francis Bacon. The affirmative side of science, consisting of its well-verified claims about the physical and biological world, may be what first springs to mind when people think about "science," but it is the critical and skeptical side of science that is the most profound, and the most intellectually subversive.

The scientific worldview inevitably comes into conflict with all nonscientific modes of thought that make purportedly factual claims about the world. And how could it be otherwise? After all, scientists are constantly subjecting their colleagues' theories to severe conceptual and empirical scrutiny. On what grounds could one reject phlogistic chemistry, the fixity of species, or Newton's particle theory of light—not to mention thousands of other plausible but wrong scientific theories—and yet accept astrology, homeopathy, or the virgin birth?

The critical thrust of science even extends beyond the factual realm, to ethics and politics. Of course, as a logical matter one cannot derive an "ought" from an "is."[49] But historically—starting in the seventeenth and eighteenth centuries in Europe and then spreading gradually to more or less the entire world—scientific skepticism has played the role of an intellectual acid, slowly dissolving the irrational beliefs that legitimated the established social order and its supposed authorities, be they the priesthood, the monarchy, the aristocracy, or allegedly superior races and social classes.[50] Four hundred years later, it seems sadly evident that this revolutionary transition from a dogmatic to an evidence-based worldview is very far from being complete.

CHAPTER 2

SCIENCE AND THE DEMOCRATIC MIND

MICHAEL J. THOMPSON

DARKNESS VISIBLE

The culture of modern societies seems, on first glance, to be imbued with science. Technological change rushes forth at an increasingly swift pace, even as advances in the hard sciences like chemistry and the biological sciences seem to be progressing according to a logic all their own. Our culture seamlessly has collapsed technology and science into a unitary reality. Science is, for far too many of us, reducible to the gadgets and mass computing systems that organize and dominate our world. Think of the technical filigree that confronts us at every turn: from how we communicate with friends and family to how global financial systems operate. But this merging of science and technology is a dangerous turn in the way that we think about our civilization, not because of some feeble Luddite appeal for social authenticity—although an argument can be made—but due to the fact that it has narrowed the actual purposes, methods, and style of thinking of science as a whole. Indeed, the erosion of science as a style of thinking, as an attitude toward the world—natural and social—is having deleterious effects on our democratic culture and on the project of a rational, democratic society.

But if we take a closer look, our society is evincing a stark decline in science as a stance in the world. In our culture, technology has become a kind of unwitting metaphor for science without most realizing that there exists a stark difference between the two, that technology is a particular application of the scientific mind-set that is applied in a narrow way to the manipulation of nature. This has done, I think, real violence to our ideas about science as a whole, for it closes off to us the idea that science can, and is, about anything other than manipulation of objects and the quantification of all things. My claim in this essay is simply this: science is a fundamental prerequisite to a rational democratic society. Science ought not to be reduced to or conflated with the features of technology and technological rationality, but rather be viewed as a means to foster reflection on objective forms of truth.

Science is a style of thinking, an attitude toward the world that grants us the power to think through our experiences of complex events, that not only fosters a sense of causal reasoning that grants us access to the mechanisms of power and constraint in the world but also cultivates an appreciation of truth as universal and objective as opposed to "truths" that are personal and particular. Even more, science is the very means by which human beings can formulate, construct, and sustain a democratic life together—one characterized by a commitment to a common life, to common purposes and goods that sustain a cooperative, interdependent society of equals. Indeed, lacking science as the basic mind-set of democratic citizenship we will be led toward a world that is narrow, fragmented, and infused by irrational beliefs, superstition, and false self-understanding, even as it continues to be suffused with technical rationality and instrumental reason. We can see the seeds of this world beginning to unfold, a kind of darkness that is becoming visible to us.

But even as the increasing dominance of technology and technological reason absorbs more and more of our culture, another force also operates against the scientific mind-set and democratic values and attitudes. This we can see in the cultivation of belief systems that seek to encourage the irrational in us, or that conform not to objective features of our lives but rather to the particular and subjective experiences we may have or the traditions and customs within which we may feel "at home." Science, for many of this persuasion, is a threat, an expression of power over one's personal choices and over one's freedoms. This can be seen in the dogmatic parent refusing to vaccinate their child or in the ideologue who refuses to transform his political ideas even in the face of persuasive evidence or in the religious "person of faith" who asserts her "truth" at the expense of all others, as well as in the university professor embracing postmodernism and the social construction of reality against the contentions of objectivism and universalism. All share, despite their obvious differences, an antipathy to science as a means of grasping an objectively true world that we all share and with which we must come to terms. Forces on both the Right and Left of the political spectrum have taken up this anti-science banner. On the Right, science is viewed as a threat to the reservoirs of tradition and authority, of religion and hierarchical relations of power in traditional forms of family life, for instance. On the Left, science is now seen as a mask for power and domination, as an "ideology" or a "social construct," a discourse like any other aspect of human culture. But this has only succeeded in highlighting the importance of the need to defend science not as a method of inquiry for the manipulation and instrumental domination of the natural world but as a *style of thinking* and an attitude toward the world that grants us access to truth-claims that are democratic insofar as they are universal, applied to all equally, and can be questioned and justified in objective terms.

But the opposition to science as a method of inquiry expresses itself as a cynicism of reason, of rationality more generally. Universals and objective truth-claims are seen by critics of science as violating particular beliefs and ideas. Science is cast as intolerant and imperious, brooking no compromise with the sensibilities of other cultures and those who choose to "think differently." But this follows from an idea about science that conflates it with the technology and the metaphor of the machine that it engenders: one that is impersonal, efficient, and oblivious to human values. This has precipitated a kind of turn against science and perpetuates the false view that science itself, as a style of thought, has been eclipsed by its own absorption into technological reason. In this sense, the social theorist and critic of science Stanley Aronowitz emphasizes a point worth taking seriously. For Aronowitz, "the statement that science and technology have become inseparable is certainly controversial, especially among those who would insist that science is autonomous from the concerns of power and ideology."[1] Aronowitz raises what we can call the *conflation fallacy*, i.e., the idea that science and technology are identical with one another or, at least, interpenetrate one another in modern society so as to be essentially unified. The aim of the criticism is that science has become an ideology of control, and the means to dominate nature as well as society. But the conflation of these two spheres is not, in the end, convincing. One reason is that we can use science as a mode of reasoning to critique technology. If we grant that technology is the rationalization of mechanisms of control, then we can certainly say that science is implicated in this, but it is nevertheless not reducible to it. Hence, professional scientists such as a J. Robert Oppenheimer or a Carl Sagan can use scientific reason to argue against the abuses of technology—of nuclear weapons, for instance, or the abuse of the environment by technical and industrial systems.

There is no doubt that the central problem with the conflation fallacy is that it forces us into a false dichotomy: if we dislike the social consequences of technology, we must also oppose science itself and view it as a discourse of power. It asks us to see science and technology as a unified historical process. But science need not be bound to the economic imperatives of its time. There is little doubt that technology and the institutions of scientific research have become increasingly fused to the apparatus of capitalism and its need for control, profit, and efficiency, but this is not the same thing as saying that science as a whole, as a mode of reasoning and as a style of thought, are also captured by this process.

The dynamics of technical change, or the harnessing of scientific efforts, are largely determined and understandable via the social aims of those who seek to benefit from them. It is not only that there is a path to dependency on technological apparatuses, but also a human interest in control. Indeed, as political scientist Kurt Jacobsen has argued, "A technology is usually selected because it suits both the criteria of technical viability and the power designs of influential groups."[2] But we can also argue that science is more than its absorption into technological logics. Science's roots in the Enlightenment were indeed contemporary with the development of modern republicanism and democratic political ideas and practices. The reason for this can be found in the new attitude that it displayed not only toward nature but also toward authority itself.[3] Once we probe this dimension of science—beyond the constrained province of the natural scientist or the technical expert—we find that science and democracy have a fruitful if not essential relationship. Science as a method of knowledge is self-correcting, just as it is anti-authoritarian; it is self-revisable and questions its own assumptions, even as technological rationality is incapable of this. Technology is self-justifying insofar as it uses its own criterion for its operation and forces us into its own logic when we interact with it. In technological rationality, logic is machine-like, there is no room for speculation, for creativity, for questioning. Everything works according to an ironclad logic that works toward a specific goal or outcome. But science as a method of knowledge is not self-validating; rather, it takes the logic of objective facts and objective reality as the only criterion for truth. Herein lies the crucial distinction between science and technology and our reason to reject the conflation fallacy.

As I see it, there is in fact a crucial, even essential relation between scientific reason—by which I mean a kind of thinking about the world that is basically rational—and what I will call the *democratic mind*, or an attitude about norms, values, and social practices that demonstrate this rationality in human affairs. I want to explore what I see to be this democratic mind and show how science is its essential, distinguishing feature. My first burden will be to explore a concept of scientific reason that is not reducible to, nor conflated with, the natural sciences, but is instead a concept of rationality that is defined by four distinct features: objectivity, skepticism, causality, and universality. Thoughts about the world and our conduct in it can be seen as rational only when these features of reason are in play. The claim I want to defend here is that this conception of rationality is rooted in a scientific mind-set and is cultivated by a scientific attitude and can be traced through the development and evolution of modern, as well as classical, forms of science and rationality.

My next task will be to show how this mind-set is inherently democratic in nature. I will do this by pointing to its anti-authoritarian impulse, its egalitarian impulse, and its transformative impulse. Taken together, these three applications of the rational-scientific form of thinking constitute a critical-democratic stance in the world by breaking down forms of thought that hold back our ability to critique, to judge, and to transform our world. We lose the ability and the style of thought necessary to be able to think democratically, that is, in objective, universal, and open-ended terms. The ability to scrutinize power, to terminate the plague of superstition and irrational traditions, and to be able to think in a way that has the common interest of society in view: these are the aims of a democratic life buttressed by the scientific attitude.

"... THEIR MINDS MUST BE IMPROVED ..."

Once we explore what the scientific mind-set actually is, then we are on our way to constructing a connection between science and democracy. Historically, there is ample evidence that advocates of modern democracy were also concerned with the idea that common citizens would be able to participate in the decision-making processes of their community. The Enlightenment ideal was expressed as the aspiration that the minds of citizens could be perfected, or at least be brought to a standard where leaders or elites could not claim power or rule as a result of a public incapacity to criticize and to question their legitimacy to hold power. Reason—the modern reason of the Enlightenment—was now allied with the impulse of democracy and the challenging of traditional authority and hierarchy. As Thomas Paine wrote in 1792, in his defense of the dawning democratic age *The Rights of Man*, "The insulted German and the enslaved Spaniard, the Russ and the Pole, are beginning to think. The present age will hereafter merit to be called the Age of Reason, and the present generation will appear to the future as the Adam of a new world."[4] Thomas Jefferson also expressed this Enlightenment principle of the political efficacy of blending science and democracy when he wrote in his *Notes on the State of Virginia*:

In every government on earth is some trace of human weakness, some germ of corruption and degeneracy, which cunning will discover, and wickedness insensibly open, cultivate, and improve. Every government degenerates when trusted to the rulers of the people alone. The people themselves therefore are its only safe depositories. And to render even them safe their minds must be improved.[5]

With this principle now in view, I want to defend the thesis of the *democratic purpose* of science. To do this, it will be important to point to the salient features of science as a way of thinking. Let's go back to my basic contention that the conflation of science and technology is a crucial error in how we view and understand what science really is. As a form of reason, science is essentially characterized by four core features. First is the idea that science addresses an *objective world* and that its claims must be *objective* in order to be valid, i.e., for them to have warrant for us to accept them as truth-claims. Objectivity is a slippery concept, but at its heart is the idea that for a notion or idea to be true in any basic sense, it must first concern an objective reality that is not compromised by our subjectivity. This should not be confused with materialism, which holds that anything true is composed of *materia*, of tangible entities (atoms, molecules, physical forces, etc.). Indeed, we can make scientific claims about more than material objects. We can say that a certain law exists, or a certain norm exists, and so on. These are not material things by any means, nor are they composed of material substrates or components. They are nevertheless objective in the sense that they are made real in the world through our acceptance of them as rules by which we all abide. This is different from, say, unicorns, which may exist in our heads but have neither objective nor material reality.

Objectivity is therefore opposed to relativism, to the notion that there can be, and that there essentially are, multiple realities or truths that are incommensurable with one another. The pursuit of science is therefore a pursuit of reasons that are independent of our subjective, particular views and ideas about how the world works. The problem of relativism is a crucial one since it opens up the possibility that reasons that cannot be scrutinized based on some rational standard. The emergence of scientific truths has generally been accompanied by disruptions in the traditional and religious belief systems of the society as a whole. The Copernican model of a heliocentric solar system, the evolutionary morphology of biological forms, and so on, all had, and in some instances continue to have, disruptive effects on traditional beliefs. But this is precisely the point: the commitment to objectivity makes the world knowable and corrects for error because our subjective ideas about the world often are tied to the traditions we have inherited as well as the errors of perception that we all make.

A second crucial feature of the scientific mind-set is that of *skepticism*. To embrace skepticism is to see any kind of truth-claim as inherently susceptible to questioning and testing. Skepticism means that nothing can be taken as legitimate simply based on its mere existence or because it is the expression or feature of an authority of some kind. Skepticism is involved whenever we look for justifications, whenever we seek to know whether that which is given to us is really what it seems to be or what others tell us it is. Skepticism is the essential nucleus of science since it is always seeking to revise the accepted truths or the accepted ideas that are used to justify and buttress our understanding of the world we inhabit. Indeed, as Carl Sagan once pointed out about the power of skepticism:

> The business of skepticism is to be dangerous. Skepticism challenges established institutions. If we teach everybody, including, say, high school students, habits of skeptical thought, they will probably not restrict their skepticism to UFOs, aspirin commercials, and 35,000-year old channelees. Maybe they'll start asking awkward questions about economic, or social, or political, or religious institutions. Perhaps they'll challenge the opinions of those in power. Then where will we be?[6]

Skepticism is therefore not only a feature of the scientific method; it also permeates outward. The key idea here is that science is not simply a method of inquiry into the natural world, it is also a means of *testing claims to knowledge*; it means producing a kind of knowledge that is rational, in the sense that is passes the test of objectivity, and can withstand the skepticism of others, in the sense that reasons are given, justifications provided.

Third, another of the central features of science is the search for *causality* as the threshold for explanatory validity. To know something in a scientific sense is to be able to grasp the causal relations and causal mechanisms that make phenomena occur. Causal reasoning is linked to the features of objectivity and skepticism I explored above, in that it forces our ideas about the world to adhere to objective processes and mechanisms that cause or in some way generate the phenomena we experience. This is a central feature of rational explanation as opposed to traditional forms of explanation based on superstition, imagination, or reasoning by analogy. Hence the relation to skepticism: once we know to look for causal claims and causal mechanisms, we become increasingly suspicious of those existent explanations that fail to make convincing causal claims. One reason for the importance of causality is the fact that the appearance of phenomena in the world and the essence of those phenomena are rarely, if ever, coincident with each another. What *appears* to us to be the movement of the sun—we still say that the sun "rises" and "sets"—is in fact anathema to the true mechanism of the earth's relation to it.

Explanation through causal mechanism thereby reveals and demystifies the experiences we have of phenomena in the world. Our modern language still retains traces of these past prescientific ways of thinking. The flu, or "influenza," is from the Italian word for "influence," since premodern European culture saw the illness as the result of influence from unknown forces or from the *malocchio*, or "evil eye," a spell cast from the jealousy of another member of the community. Of course, knowledge of the objective causal mechanism of the illness yields a very different explanation, one that demystifies the world, one that allows us to grasp the real cause and force behind the phenomenon. We now have the capacity to act to prevent or minimize the spread of the illness as a result. We can see the same thing in the social and political world. Patterns of income inequality or of educational outcomes and so on all have causes that are social and structural in nature. Generally speaking, when we lack causal reasoning skills, we tend to fall back on ad hoc and traditional ideas to "explain" these phenomena. Perhaps equally as bad, when we lack causal reasoning skills as well as a strong sense of skepticism, we may be seduced into pseudoscientific arguments about these phenomena, such as genetic explanations, or arguments about the logics of markets, and so on. Whatever the case, causality and sensitivity to causal reasoning is a central aspect to a scientific and rational comprehension of the world and our experiences of it.

Lastly, there is the importance of *universalism* for any kind of scientific reasoning and criterion for acceptable knowledge. In its most orthodox form, we see this whenever we assert that $2 + 2 = 4$ and consider the statement true because of its universalism, i.e., that $2 + 2$ can never equal 3 or 5 or any number other than 4. If it did, and if it could, then the very operation of addition would cease to be an actual process. It would be essentially meaningless, nonsense. In the same fashion, whenever we define a triangle as a three-sided shape with the sum of the interior angles being 90 degrees, we are making a universal claim about the property of triangles. In other words, triangles are not conceivable or rationally knowable outside of this universal feature. What these two examples illustrate is the feature of universalism as an attribute of a scientific claim. For something to be rational it must indeed, in some basic sense, be *universalizable*. If it lacks this quality, it cannot be a truth-claim. Even in the weaker claims about social facts or morals we cannot say that "all citizens have a right to X" and then deny the right to X to some members of the community based on some arbitrary or ascriptive trait. Once we are able to spot contradictions within any proposition and its application we are able to critique that proposition as irrational. The move toward a more democratic society is premised on this as a basic criterion: that the norms and ideas that guide our social relations and institutions must be universal, applicable to all, or at least possess or instantiate a principle that is applicable to all. Universalism is a feature of a reason that ought to count, that ought to be taken as asserting a valid truth-claim or proposition.

These four features of scientific thinking are non-reducible to technological logic. We need not commit ourselves to the conflation fallacy and assume that science and technology are collapsed into a unified social construction. Rather, these four features of science capture the essence of a kind of rationalism that can be used not only for the understanding of the natural world but also the social world. But unlike technology, this knowledge need not be *instrumental* in nature; in other words, this kind of reasoning and thinking does not need to be tethered to an interest in manipulating and controlling the natural or social world. The tendency of scientific institutions and efforts to be harnessed by social forces—of military and corporate interests, for example—is no fault of science *per se*. It is rather the result of the imperatives of elites who orient and shape the practices of science and technology toward their own interests and purposes. This is the great fear that motivates some critics of science: that it has within it an intrinsic logic toward instrumental reason, or a kind of rationality that is defined by a single goal or outcome, and is really only an expression of social power. But the four features of scientific thinking that I have pointed to above are not reducible to instrumental reason nor do they play into the hands of the powerful. Rather, what seems to be most relevant here is the capacity to generate forms of knowledge that can be tied to a different set of values other than profit, efficiency, social control, or destructive power. What I have in mind here are *democratic* values, and these values and features of democracy should be seen as essentially coexistent with the scientific mode of thinking.

AN ANATOMY OF THE DEMOCRATIC MIND

If the scientific mind-set, as I have outlined it above, is to have any importance for democratic life and modern citizenship then we must see how these ideas about objectivity, causality, and universality are to be translated into a democratic and political context. Science, as I have been framing it here, sets up for us an objective context within which claims to truth and claims to valid arguments can be asserted, defended, and accepted. This has become the focus of many thinkers in contemporary political philosophy. The twentieth century witnessed a turn toward understanding democracy as deliberative, as a set of practices that allow citizens to exchange reasons and justifications for their beliefs or ideas about moral and political affairs. The problem with this academic trend is that it generally overlooked the fact that the actual institutions, culture, and social order of modern capitalist societies do not demonstrate or encourage the kind of scientific mindset required for such practices to produce rational, democratic outcomes. When we consider the relation between science and democracy, we must therefore move beyond the idea that science is solely a method and one that can be sustained within a social context that is itself irrational without some sense of conflict and struggle.

The relation between science and democracy is a root problem because in many ways democracy is a problem of knowledge. To be a democrat (in the general sense not specific to a particular political party) it is not necessary to be a scientist, but it is necessary to think a certain way, to think about the social world not in narrow, *personal* terms but in objective and *common* terms. This is a change in thinking style, in the attitudes that underwrite our thinking, not a more demanding or unrealistic epistemological level of thinking. Objectivity as opposed to subjectivity as a style of thought is a great challenge for modern, postindustrial cultures, particularly because of their greater emphasis on personal experience and self-expression as cultural ideals. But, as the Swedish economist and sociologist Gunnar Myrdal rightly pointed out, "Ignorance, like knowledge, is purposefully directed. An emotional load of valuation conflicts presses for rationalization, creating blindness in some spots, stimulating an urge for knowledge at others, and, in general, causing conceptions of reality to deviate from truth in determined directions."[7] What this means is that objectivity—one of the core features of the scientific mind-set and an attitude that science cultivates within our thinking—is directly related to the ways that we relate to the world: how we understand social problems, social power, and other features of our community. Relating science and democracy is therefore not a matter of training people to be scientists, it is a matter of holding the ideas of citizens and institutions to a standard of reason.

When Jefferson argued that for a democratic society to work its members' "minds must be improved," it necessitated the cultivation of a scientific way of thinking that was rooted in the institutions and education and practices of the society as a whole. After I explore what I see to be the basic features of the democratic mind, I will return to this problem, since it is perhaps the most important problem facing modern democratic societies. The features of the democratic mind that I would like to highlight derive from the core features of the scientific mind-set I sketched above. First, there is its *anti-authoritarian* impulse, the idea that all forms of authority and power are to be shared in the sense that each of us grants legitimate authority to others or to institutions based on our rational consent and that any claim to authority is not accepted without rational reasons. Second is the openness to either question power or to submit to it based on this criterion of *rational legitimacy* and acceptance. The democratic mind knows how and when, to use Aristotle's phrase, "to rule and be ruled," i.e., knows when to follow the better idea or principle and when to seek to propose one that is better for all concerned. Last, I want to point to a desire for *universalism*. The democratic mind and democratic attitude is one that embraces what is good for the whole community and what is right, should be right for all.

Taken together, these three features of the democratic mind possess radical implications and they are rooted, in an essential way, in the scientific mind-set itself. Let's examine each of these features of the democratic mind in turn and explore how they, taken together, are not only rooted in the scientific mind-set but also help us get a clearer view of democracy as a political form. Take the first feature of the democratic mind, what we can call its anti-authoritarian stance. Any scientific attitude must be skeptical of authority—that is, skeptical of the prevailing ideas that are put forth as valid and legitimate. Immanuel Kant made this a central idea in his epistemological theory of rationality. For him, reason and freedom were both mutually necessary concepts: rationality was in fact dependent on our ability to use our reason freely. A mind that was dependent on authority—ecclesiastical, political, moral, technical, or whatever—was essentially incapable of reason since reason itself depended on the capacity to question the constraints of tradition that is exerted on thought. "Freedom," he wrote in the preface to his *Critique of Practical Reason*, "constitutes the keystone of the whole structure of a system of pure reason,"[8] and this expressed the view that for the mind to be rational it had to wrest itself from the external authorities on which it depended.

This had not only epistemological but also moral and political implications that are inherently democratic. We not only resist authority for the libertarian sake of resisting authority, we resist authorities that cannot justify a law, a norm, an idea, or proposition. For reason to be possible, we need to be *free*: we need to be free from our dependence on ideas and ways of thinking that are not our own, that are imposed on us by tradition, by fear, or by conformity to the ideas of the community.

This idea has a strong relation to the next feature of the democratic mind. Aristotle—no great friend of democracy as a pure system of government—nevertheless argued that a central feature of how free citizens govern themselves must be found in a principle that any reasoned citizen would follow: that of the ability to "rule and be ruled." As he puts it in his *Politics*:

> There is a rule . . . that is exercised over free men and equals by birth, a constitutional rule, which the ruler must learn by obeying, as he would learn the duties of a general of cavalry by being under the orders of a general of cavalry, or the duties of a general of infantry by being under the orders of a general of infantry, and by having had the command of a regiment and of a company. It has been well said that he who has never learned to obey cannot be a good commander. The excellence (*aretē*) is not the same, but the good citizen should be capable of both—he should know how to govern like a free man, and how to obey like a free man. These are the excellences of a citizen.[9]

Aristotle's basic principle here is predicated on the assumption that free people follow reason, that in order for them to realize and keep the common interest of the community in view, each would need to be able to accept the rule of the best idea, the best reason, and perhaps even the best person. This was not a dogmatic, passive acceptance of authority, but was rather a means of generating agreement on forms of cooperation that are intrinsic to political life. True democracy is not only a challenge to authority *because* it is authority; it must also be a search for legitimate, rational forms of authority that can serve common interests with which each member of society can identify.

In this sense, the democratic mind is not only resistant to any passive acceptance of authority, it is willing to accept as legitimate only those ideas that are justified based on the inspection of reason. This reason becomes democratic not from our mere agreement, but once it can be demonstrated that a given norm or institution satisfies the common interest of its members. If we combine Aristotle's thesis with the Enlightenment, scientific idea of anti-authority in knowledge and moral-political life, then we can see that a more democratic principle begins to emerge: an openness to the most rational ideas that are best for our living together in a rational, self-governing community. To privilege tradition, to make a claim for the exception of any individual or group, or to accept passively the ideas, claims, norms, and justifications of those in power or that which merely exists, is a decidedly *anti-democratic* kind of thinking and attitude. It is also, as I have been trying to show, deeply *anti-scientific*. The democratic mind is not a closed mind, it is not committed to only one principle—say the narrow libertarian idea of being against *all* authority—it is an *open* mind: one that will revise its ideas, values, and norms based on new evidence and more compelling reasons. Reason therefore does not only challenge ossified forms of power, it also can provide the criteria for new forms of life, more democratic, egalitarian, inclusive forms of life and politics.

Once we consider the problem of an open and a closed mind, we come to the problem of universalism versus particularism in human affairs. Basically stated, universalism in moral terms means that there are certain values that apply to all persons. Cultures can therefore be judged according to their ability to engender values that respect its members based on universal, objective criteria instead of their own particularist traditions, values, and ideas. The problem here is not only a meta-problem, it is also, and perhaps more centrally, a problem of how an individual thinks about the world, his community, and the fabric of norms and values that pervade them both. Seen in this way, the scientific mind and the democratic mind find an special resonance with one another. The main reason for this is that our knowledge about things is highly dependent on the values we carry with us. As social psychologist Milton Rokeach points out:

Every person may be assumed to have formed early in life some set of beliefs about the world he lives in, the validity of which he does not question and, in the ordinary course of events, is not prepared to question. Such beliefs are unstated but basic. It is out of some such set of "pre-ideological" primitive beliefs that the total belief-disbelief system grows.[10]

The values that we learn early in life and through exposure to the social and cultural norms and traditions around us shape us and form our ideas about the world—not only our normative ideas, about what the world *ought* to be, but our descriptive ideas as well, i.e., our ideas about what the world *actually is*. Our beliefs about the world, embedded deep in our psyche, therefore are a constraint on our ability to think objectively and scientifically about it.

This idea is an important one since perhaps one of the most powerful implications of the scientific mind-set and its relation to the democratic mind is its ability to break down these latent values and beliefs. Greek philosophy, one of the first expressions of a scientific mind-set and its relation to practical questions, was also deeply committed to this view. Plato—no advocate of democracy, to be sure—was nevertheless prescient about reason's ability to move our ideas away from subjective error toward objective claims. In the *Republic* he refers to this as "dialectic" and describes it as a process of self-transformation through the purging of all presupposed beliefs:

> The dialectical method is the only one which in its determination to make itself secure proceeds by this route: doing away with its assumptions until it reaches the first principle itself. Dialectic finds the eye of the soul buried firmly in a kind of morass of philistinism. Gently it pulls it free and leads it upward, using the disciplines we have described as its allies and assistants in the process of conversion.[11]

Plato's idea here is a powerful one. Dialectic is a kind of reasoning that breaks down the assumptions and presuppositions that affect knowledge and bias our ideas about the world. It is a process that constitutes self-transformation. The thesis is that reason's most powerful, perhaps highest role, is to dissolve our assumptions and presuppositions that prefigure our knowledge about the world. This is one way of thinking about the capacity of science to equip the democratic mind with the need to search for universals, or what Plato calls "first principles." These universal values and norms transcend the particular views that we carry with us and to which we are, more often than not, not only cognitively and morally committed, but emotionally committed as well. It is perhaps for this reason that, in his construction of democratic personhood, Jean-Jacques Rousseau remarks on the importance of truth as impersonal and objective: "I know only that truth is in things and not in the mind that judges them, and that the less of myself I put in the judgments I make, the more sure I am of approaching the truth."[12]

THE CULTURE OF ANTI-SCIENCE AND THE FATE OF MODERN DEMOCRACY

If my argument above has any meaning today, it can be glimpsed in the disturbing trends toward anti-science that grip modern societies. The fate of modern democracy is, I assert, endangered to the extent that the scientific mind-set and the democratic mind do not reinforce one another. Today, a culture of anti-science is growing in its reach and in its influence. There is an increasing ignorance on the part of the public of the essential components of scientific reasoning, even as there is a growth in the technical apparatus of modern societies. But the attitudes and thinking of the public have been deleteriously affected by the growing influence of anti-science viewpoints and anti-rationalist beliefs about the natural and social world. Any time we encounter denial about human-made climate change, the belief in racial causes for social pathologies, religious denial of scientific fact such as in evolution or cosmology, or the legitimacy of conspiracy theories, we are glimpsing the face of anti-science.

Many contemporary critics of science often point to the cultural diversity of ideas and beliefs and to the richness and value of this diversity of moral, religious, and knowledge systems, arguing that the acceptance of these diverse ideas is itself a democratic value. "A free society," writes philosopher Paul Feyerabend, "is a society in which all traditions have equal rights and equal access to the centers of power."[13] But we should examine this claim carefully. The relativists' challenge to universalism works only insofar as the society within which it operates has fragmented to the point where common ends and purposes have lost their normative legitimacy. I can claim my own truth, in this sense, only when there are no shared truths in which the society I live in is invested. When we give up on the objective worldview and move away from universalism as a category of validity for our ideas about the world, we lose the capacity to critique the forms of power that our social world manifests. If we assert the primacy of our traditions, we lose sight of the biases of those traditions. The universals that we come to accept via our rational faculties are foundations for the kinds of ethical life and institutions we find valid. The democratic mind is one that is able to move outside of egoistic interests alone, one that is able to transcend the biases of one's cultural and traditional belief systems, and one that places emphasis on the goods, ends, and purposes that are shared in common with others. The scientific mind-set is therefore the undergirding style of thinking that grants the democratic mind its shape and its function. Indeed, it is an essential feature of a democratic culture and polity.

Science as a mind-set and as an attitude about objectivity, universalism, causality, and skepticism is therefore more than a methodology for the practice of scientific investigation. This attitude is not empty of value; it is in fact value-laden once we consider that it is an attitude about how we should relate to knowledge claims as well as to the world in general. Science is not neutral to democratic practice and a democratic style of thinking. Indeed, science may not be able to provide us with the *content* of our moral concepts, but it is necessary for a democratic ethical life and polity. Max Weber's thesis that science and philosophy need to be kept separate cannot be maintained since science is not merely a formal method but also an *attitude* and *orientation* that the mind takes in its relation to the world.[14] In this sense, the boundary between science and democracy begins to break down. Once we see democracy merely as a competition between different interests and worldviews, we have committed ourselves to an irrational view of democracy—indeed, to a pre-modern conception of democracy. The modern theory of democracy, premised on the Enlightenment conception of the democratic citizen as possessing rational agency, an interest in the common good, and an antipathy to irrational authority (oligarchic, tyrannical, or populist) is therefore equivalent to the scientific attitude.

Leo Tolstoy once remarked that science is meaningless "because it gives no answer to our question, the only question important for us: 'What shall we do and how shall we live?'"[15] But even though we can agree with the claim that science cannot provide us with the values for such normative questions, we can still say that science—as I have been describing it here—provides us with values in the sense that it cultivates an attitude with respect to authority, to valid knowledge, and to our relation to our and others' beliefs. Science does not provide us with values, but it embodies a set of attitudes toward the world and, in so doing, can be said to manifest values of universalism, objectivity, and inquiry. These may not be substantive values, but they are values nevertheless in the sense that they provide us with a preference for skepticism over faith, for universalism over particularism, and with objective facts rather than subjective preferences and predilections. The scientific attitude and mind-set is itself a commitment to a set of values about how we should relate to evidence, to reality, to decision-making, and so on. This does not mean that we obliterate our values or that we suppress our interests; it does not mean that we erase our subjectivity or our agency. It does mean, however, that in democratic affairs, we must be willing to leave our purely personal perspective and think in a more expanded, objective way in light of more rational, more universal interests—that our subjectivity should change in its relation to our rational comprehension of objectivity and that our agency should be shaped by our cognitive powers to inquire and to transform our thinking in accordance with new, more compelling truths. What it means, in the end, is that we relate to our values and interests *rationally*, i.e., with a critical attitude that will either force us to revise them in light of better reasons or to defend them in universal terms that others within the community can accept as democratic.

Although this move toward a critical rationalism based in science was the hallmark of modern political thought, the appeal of the anti-science position does not seem to be diminishing. In postindustrial societies in particular, the emphasis on identity, self-expression, and personal experience has fragmented the cultural and political terrain. With the embracing of anti-science and anti-rationalist positions, a celebration of the particular now dominates much of contemporary thought. Stanley Aronowitz argues for "the rejection of universal reason as a foundation for human affairs." As he elaborates this position:

> Reason in this sense is a series of rules of thought that any ideal, rational person might adopt if his/her purpose was to achieve propositions of universal validity. Postmodern thought, on the contrary, is bound to discourse, literally narratives about the world that are admittedly partial. Indeed, one of the crucial features of discourse is the intimate tie between knowledge and interest, the latter being understood as a "standpoint" from which to grasp "reality."[16]

So, for Aronowitz, the anti-science position is beneficial since it gives precedence to the identity of individuals and groups by undermining the theory of objective knowledge. What results is a proliferation of claims to particularist knowledge(s) rather than knowledge that can pass the test of reason thereby damaging democratic culture and thinking. This position only contributes to a culture that eschews rationality and the need to transcend the narrow confines of tradition, belief, and identity in order to critique social relations of power, control, and domination. Such a particularist and perspectival position can only lead to the shredding of political solidarity and a moral narcissism that fragments modern democratic life. Power can, and indeed must, be critiqued and questioned without the assumption of anti-scientific and anti-realist views.

Add to this the way that the shredding of scientific reason and respect for evidence has strengthened anti-democratic policies on the environment, workplace safety, drug policy, and other aspects of modern life. It has done this because, without a respect for objective facts and evidence, we fall back on alternate forms of authority for our knowledge claims and beliefs. We may fall back on our common experience, our traditions, or on religion or some other form of custom or authority that will give warrant to our beliefs and claims. But in doing this, we also fragment the very preconditions for a rational public sphere, for a set of attitudes, practices, and cognitive norms that allow for a democratic expansion of knowledge necessary for any community to govern itself rationally. Lacking this, our attachment to external and irrational authority, tradition, or "common sense" and "experience," undermines our ability to exert democratic control over elites who control corporations and government, as well as democratic control over technological change, and the increasing powers of economy and the state. Disrespect for objective evidence, an antipathy toward the skepticism of views that resonate with one's intrinsic values and biases, is not a democratic defense of difference and the legitimacy of different traditions. A democratic society consists of members who rationally reflect on all forms of power, custom, and tradition and seek binding agreements on norms and institutions such that they fulfill the common good of its members. Democracy is only valuable to the extent that it fosters a self-critical attitude among all of its members.

NIHILISM AND UTOPIA

There is an ironic twist to the argument I have laid out in this essay. One of the core impulses of the critical reaction to science throughout the nineteenth and twentieth centuries has been the charge of nihilism: that science would bleed ethical values and aesthetic sensibility from human life. Science was a quantifying, narrowing form of thinking that would squeeze out moral judgment from human affairs and serve as a procrustean bed on which all knowledge would be flattened. But as I have tried to show here, this charge—although not without merit—is misdirected about science and its relation to democracy. The main reason is that once we reject the conflation fallacy of the intrinsic merging of science and technology, we see that science is not an inherently instrumental attitude or mind-set. Rather, science asks us to think in a framework where only *rational reasons* count—and rational reasons have as their basic criterion of validity their claims to universality. This does not in any way invalidate our desire for toleration of different views or opinions as to certain moral choices. But it does invalidate the idea that there can be multiple truth-claims and that a hierarchy of what we deem valid can be established.

The problem of contemporary culture seems to be one where common values and assumptions about our institutions and the aims and purposes of our social world are evaporating. Much of this can be blamed on a culture that has embraced technique at the expense of the ends toward which our society is oriented. The idea that there are common social purposes and ends, indeed, a common good toward which our political, economic, and cultural institutions ought to be oriented has now disintegrated under the auspices of a concept of society that is highly atomistic and alienating. The forces of technology and unequal economic power continue to break down the civic life of modern society, but it is also evident that the spread of the technification of society—of our social relations, our work, and our daily interactions—has privileged means over ends. Disappearing are the spontaneous aspects of science: of experiment, of creativity and curiosity in the new and the impulse to know the unknown. We are becoming rigidly circumscribed by our conformity to technological imperatives. Technical reason is now becoming hegemonic over our rational abilities to control and direct it according to common goods and ends. Technology is not science—and it is not democratic. What is needed is the reconstitution a kind of rational democratic society that can orient technological change according to democratic ends.

The exacting nature of technology as a means of control—expressed as efficiency and predictability—has eroded our moral-reflective powers on the proper ends of our institutions and social roles.[17] As such, technology and science are becoming increasingly a mechanism of control and of profit, absorbed into the cybernetic logic of technologized bureaucratic capitalism. Technological management in working life, education, and culture now alienate us even further from the concept of a society organized around a common good. The result is a stark nihilism: a reified form of consciousness where our concept of other people, ourselves, and the natural world are becoming increasingly the objects of manipulation and control. The breakdown of codified traditional belief systems through the course of the nineteenth and twentieth centuries has meant that a world that is calculable, predictable, and hence controllable has become more attractive to us. As Neil Postman correctly argued on this point, "The thrust of a century of scholarship had the effect of making us lose confidence in our belief systems and therefore in ourselves. Amid the conceptual debris, there remained one sure thing to believe in—technology."[18] The results of this nihilism have been severe. On the one hand, we cede our moral and critical capacities to technical systems, while on the other, we react to our social meaninglessness and move into the realm of tradition, religion, custom, and fabricated identities. We lack a common value system that can deal with these concerns and we seem increasingly to be giving ourselves over to the cybernetic logic of a society imbued by technology. A science and technology produced by and embedded in capitalist imperatives for profit and accumulation can only continue to grind down the autonomous self just as it erodes a value system that can give shape to a truly democratic and rational form of life.

The alternative impulse in modern society is toward what we could call a utopian idea about the future of technology and the application of science. According to this view, which is fundamentally ideological, technological change has its own logic that must be allowed to unfold. This applies to ideas not only about applied technologies in electronics, computing, artificial intelligence, and so on, it also applies to ideas our society tends to hold about the economy, "markets," and trade. These entities are granted a status that likens them to organisms or autonomous systems, properties of nature rather than products of human artifice. But this is actually not the result of the spread of science but its narrowed use and capture by elites. Once we commit ourselves to the view that science is at its base a mode of inquiry, of thinking, an orientation toward the world—once we accept that what I have been arguing here is the intrinsic relation between science and democratic thinking, only then will we be able to explode the ways that science and technology have been and continue to be employed not for human betterment and emancipation, but rather for the expediency and interests of the powerful. Only then will we be able to counter the utopian ideology of technology and science and crack open the anti-democratic and hierarchical social structures the constrain a truly free, genuinely democratic society.

Nihilism and utopia are therefore two mutually supporting attitudes that pose real threats to the culture and politics of a democratic society. For both engender an anti-democratic worldview insofar as they surrender their powers of self-governing to some external authority or system. Both encourage a conformity to views and practices and norms that lure us into the technical systems of modernity. But how can we not mourn the loss of the autonomous self? How can we not avoid confronting the de-democratizing tendencies in modern culture no less than its de-rationalizing currents? Science needs to be rehabilitated, not as a mere tool for the domination of nature and the manipulation of society and individuals, but as a means of rational reflection and the examination of the ends and purposes toward which we orient our society. The continuing impact of technological control on modern civilization is such that it continually robs the individual of the capacities for cognitive reflection and the autonomous ability for critical judgment. Science, as I have described it in this essay, must therefore be rehabilitated in light of its democratic potential. Indeed, it must be rehabilitated as the very framework of rationality itself, as the surest means to secure our knowledge about the world. Both science and democracy are therefore ways of knowing and ways of acting. Only a political project that sees their mutual nurturing of each other will be able to preserve the hope and the vision for a free and rational society.

CHAPTER 3

THE SYNTHESIS OF SCIENCE AND DEMOCRACY: A DEWEYAN APPRAISAL

JOSEPH CHUMAN

THE RISE OF ILLIBERALISM

With the fall of the Soviet Union, it was widely assumed that liberal democracy would spread and endure, enjoyed by increasing numbers of people on all continents. The end of communism and the democratization of authoritarian states, as well as the unprecedented economic uplifting of hundreds of millions of people who had lived in destitution, all spoke to the triumph of liberal values. It was a time when faith in the Enlightenment could be renewed: Democracy, economic development, human rights, scientific and technological wizardly, expressed most dramatically in the growth of digital technologies, emerged to usher in the new millennium. All spoke to an upward arc of continuing progress.

Yet in a few brief decades this optimism has been severely challenged. In rapid order we have witnessed nothing less than a fundamental disenchantment, not only with democratic forms of government but also with the very foundations of science and intellectual integrity on which the modern, democratic state rests. The emergence of "illiberalism" is widespread and very ominous. In its political manifestations we see the emergence of nationalist parties and fervent anti-immigrant backlashes in Western Europe, growing authoritarianism in Eastern European states, as well as in Turkey, China, India, and elsewhere. And most strikingly, we experience it in the United States in ways that bring into question, as never before, the stability of our democratic institutions.

This new nationalism expresses itself in populist movements and attacks on a free press and an independent judiciary. In a manner frighteningly reminiscent of mid-twentieth century movements, one hears of a yearning for "strong men," as faith in the responsiveness of democratic governments grows weaker.

Despite the optimism felt after the Soviet period, there were earlier signs of dark times brewing on the horizon. The emergence of religious nationalism and political movements justified by religion—most dramatically represented by the Iranian Revolution of 1979—huge and growing economic disparities intensified by the triumph of neo-liberalism, and the persistence of mass atrocity crimes, all spoke to the intensification of parochial loyalties out of despair with universal values and a breakdown of wider allegiances characteristic of the modern liberal state.

THE AMERICAN CONTEXT

What is occurring across large sectors of the international arena, we are experiencing in the United States in unprecedented ways. Dark forces have widened and metastasized to challenge not only democratic institutions, but also the most fundamental norms that make a coherent society possible.

These forces of illiberalism are dramatically expressed through the persona and policies of Donald Trump, who defied the polls and the pundits to win the White House in 2016. But Trump represents the culmination of trends that began much earlier.[1]

The surprising presidential victory of Trump in 2016 is an exemplification of economic, political, and intellectual trends that have long been brewing. The result is greater warrant at this moment to fear for the future of American democracy than at any time since the Civil War. Our democratic institutions, which Americans have long taken for granted—the independence of the courts and the media, our system of checks and balances, fundamental respect for our Constitution—are now being undermined. But even more menacing, as noted, is the gratuitous assault on the very norms that underlie democratic institutions, and beyond, the very fabric of social life. Among them are such intuitively basic foundations as the appeal to facts, truthfulness, and rational consistency, and the employment of the rules of evidence as components of rational decision-making. Among the leading casualties of this ominous trend is a cavalier and politicized disregard for the authority of science. President Trump provides the leadership and legitimation of the these subterranean forces, but he won the White House by appealing to a grassroots constituency of prevailing working-class voters who feel left behind as America has shifted to a postindustrial economy, and an influx of foreigners have created a cultural landscape and society that they sense is increasingly alien.

These changes have generated an ugly anti-immigrant backlash and have unleashed racist and white supremacist outbursts unwitnessed on the national level in decades. We now live in a divided society. Partisan political divisions have resulted in a stagnated Congress. At the popular level, those who feel themselves disadvantaged and who are apprehensive and angry that the doors of economic opportunity are shut, especially for the next generation, view those living on the coasts as "liberal elites" who hold with contempt those in the American heartland. A result is the emergence of a new tribalism, centered not so much around ethnicity and religion, though that endures, as much as political ideology and one's position on defining issues such as abortion, gun ownership, gay rights, government regulations, and other causes that have been the centerpieces of the culture wars spawned in the 1960s. A searing causality of this tribalism that has excoriated putative elites and their values is an assault on higher education and the knowledge and authority that education brings. Absorbed into this assault is profound anti-intellectualism that animates a distrust of science and, beyond, the compelling nature of facts and evidence, when science, facts, and evidence threaten tribal identity and cohesion.

Without a consensual commitment to these baseline values, we have entered a surreal universe and have opened the door to societal breakdown. It is against the background of these startling and ominous developments that those committed to the maintenance of a viable American future need to forcefully advocate for democracy and engage in a renewed process of education toward that end. Even after Donald Trump exits the scene, there will be much rebuilding to do.

Science and democracy, both under attack, are the twin pillars of modern society, and, for those of liberal sensibilities, they are mutually sustaining. While one can imagine scientific endeavor taking place in nondemocratic environments, and with greater difficulty democracies wherein scientific progress is not given much emphasis, such a scenario contradicts the unfolding American narrative that has defined the country's self-understanding since its founding. An American society that is not democratic and that depreciates the importance of science would be a betrayal of America's core identity. It would also result in a much diminished America and the country's retreat from global leadership, where it has maintained a prominent position since World War II.

Whether the Trump phenomenon is transient or opens the door to an even darker future is impossible to predict. But a strong dynamic and counternarrative coursing through the American character is an appreciation that the future is open, and that it can be guided by the values and visions of its people. If this is the case, then the future calls for a renewed commitment to democracy and a reclamation of the importance of science and the scientific spirit.

The task is to get beyond stridently polarizing tribalism and inspire a reinvigorated enchantment with democratic values and processes. Solutions, needless to say, will be complex and long-range. A restructuring of the economy will be required, in order to reopen doors of opportunity that are currently closed and restore upward mobility, which is stalled. This will include radically reducing the unprecedented wealth gap that weakens democracy and democratic participation. As a broad generalization, when reality no longer works for people, irrational explanations for their condition become more attractive. Ways must to found to take big money out of politics. Reinvigorated moral leadership needs to emerge to re-inspire a divided society with unifying visions of the American Idea beyond racism and xenophobia, speaking to the strengths of an inclusive pluralism. What is also needed, as noted, is a reacquaintance and re-enchantment with the principles of democracy and civic involvement at all levels, and a deepening understanding that these principles are not only intrinsic to the American way of life as we have known it, but that they hold out the promise of the best life for all.

THE RANGE OF DEWEY'S THOUGHT

The times we are in, therefore, make a renewed appreciation for the thought of John Dewey especially relevant. Dewey's was arguably America's premier philosopher of democracy and a staunch defender and expositor of the scientific outlook. In briefest terms, Dewey, more than perhaps any other modern thinker, created a fusion between the democratic process and the scientific method. For Dewey, the scientific method entailed a democratic process and, similarly, democracy encompassed a process that is broadly scientific. For Dewey, "Democracy as a way of social life is the expression of a humanized experimentalism. In other words, the ethics of creative democracy are an expression of wide sympathy and a heartfelt concern for the common good, guided by experimental empiricism."[2]

Dewey's philosophical output was prodigious and ranged very broadly. "He worked philosophically and at great length on problems of ethics, art, psychology, sociology, logic, religion, politics, and the area for which he is best known and was most influential—education."[3]

Concepts as such as "democracy," "science," "empiricism," "experimentalism," and "the common good" are centerpieces of Dewey's philosophy and point to both his method and the aims toward which his entire philosophical output was directed.

Dewey was born in Burlington, Vermont, in 1859, completed his doctoral work at Johns Hopkins, and started teaching at the college level at the University of Michigan in 1884. Ten years later, he moved to the University of Chicago, where he assumed the chairmanship of the philosophy department. Dewey's move away from purely speculative issues accelerated during his time in Chicago, in part through his acquaintance with Jane Addams, the famed social worker and founder of Hull House. It was at Hull House that Dewey became directly involved with the problems wrought by urbanization, social change fueled by industrialization, and the plight of immigrants. It was here that Dewey associated with workers, union organizers, and radicals of all stripes. He was most impressed that the residents, in accordance with Addams's philosophy, were not the objects of a patronizing charity but were self-governing in working through the challenges they confronted. There can be little doubt that these early experience were formative in molding Dewey's philosophy of democracy.

It was also in Chicago that Dewey founded the Laboratory School at which he could test out his pedagogical hypotheses. This was the period in which Dewey developed the philosophy of education for which he best known and developed an international reputation.

In 1904 Dewey arrived at Columbia University and became the luminary of its philosophy department in what was arguably the department's Golden Age. It was here that he expounded on his philosophical naturalism and its associated applications to a wide ranges of disciplines. Dewey continued with his life-long commitment to activism, even while becoming America's premier public intellectual. He marched with teachers on picket lines in support of their unionization. When he was seventy-eight, he traveled to Mexico City to lead a commission investigating the charges brought against Leon Trotsky, who had been condemned in the trumped up Moscow trials. And he supported and provided hospitality to Bertrand Russell when Russell was denied a teaching position at City College on the grounds of being moral unsuitable. It is noteworthy that Dewey wanted to ensure fairness for Trotsky despite believing him to be a fanatic, however brilliant. And Russell was a philosophical adversary, who had been less than charitable to Dewey. Perhaps most relevant to our interests here was the fact that Dewey's personal values and his public commitments expressed the engagement that lay at the heart of his technical philosophy.

Despite the extensive scope of his thought, it would be correct to conclude that Dewey's major philosophical project was to break down dualisms that plague both technical philosophizing and ordinary modes of thinking.

> Dewey's grand intellectual effort was to show the harmony and continuum between ideas and action, thinking and doing, science and religion, science and art, means and ends, facts and values, ideals and reality and . . . the interrelatedness of the individual and the community. Classical philosophy usually emphasized the separateness of these categories. Dewey, inspired in his youth by the philosophy of Hegel, sought to show their similarities and connections with each other.[4]

While democracy and science are ostensibly different endeavors, their overlapping characteristics place them within Dewey's grand project of harmonizing if not unifying ostensibly disparate aspects of thought and behavior. In regard to science, Dewey is not primarily interested in the specifics of scientific discovery. And when he discusses democracy, he does not focus on the institutions required for democratic governance, though he affirms these as essential. Dewey, in concert with his naturalistic metaphysics, posits that nature and reality are continually in flux.[5] Dewey began his career as an Hegelian idealist, but his later thought expresses an explicit break and contrast with idealistic metaphysics. All things change, and he rejected fixed points and absolutes. As such, when it came to both science and democracy, Dewey assessed them primarily as *methods*. This close kinship of democracy and science will become clearer by discussing in some detail what he means by each.

DEWEY AND SCIENCE

For Dewey, scientific thinking was a refinement of the processes of thinking in general. In Dewey's view, reality is inherently problematic. In simplest terms, as we move through life, we confront problems at every turn. We are impelled to resolve each problem in ways that contribute to the greatest fulfillment. Once resolved, we then move on to engagement with the next problem. It is the method of resolution that most interested Dewey, and stood at the center of his philosophical pragmatism, which he also referred to at various times as "instrumentalism" and "experimentalism."

Dewey stands as one of the major formulators, together with Charles Sanders Peirce and William James, of pragmatism, which has often been characterized as a distinctively American philosophy. In its theory of truth, pragmatism broadly identifies what is true with what "works" with regard to creating human satisfaction. In this sense it contrasts with forms of idealism that measure truth by how well the proposition or hypothesis at hand is consistent with an objective, independent, and preexistent reality. Pragmatism extends beyond the resolution of practical problems to embrace moral ones as well. This implies, again, that moral values do not derive from absolutes but emerge out of concrete circumstances and frustrations that seek satisfactory resolution.

Steven Rockefeller provides a summary of how this process functions in Dewey's thought:

> Instead of seeking for absolute ideals and offering ready-made solutions to moral problems, pragmatism adopts a genetic and experimental approach to moral problems and focuses on developing a method for dealing with specific moral difficulties as they arise in concrete situations. It directs a person facing a moral dilemma to carefully clarify the nature of the problem and then to give attention to specific alternative values or ideals that might guide conduct in the situation, noting especially the conditions that might actualize them, that is, the means to their realization. With the aid of this knowledge of conditions or means, it studies the actual consequences that flow from acting under the guidance of the alternative values in question. In the light of a knowledge of consequences, it then evaluates these ideals, taking into consideration the specific needs of the moral problem at hand. This is the process of pragmatic moral evaluation.[6]

Central to Dewey's theory of knowledge is the employment of "intelligence" as an essential component of inquiry, which Dewey never tired of invoking. Intelligence is contrasted with "reason," which Dewey interpreted as the assertion of prior, fixed absolutes that stand outside of experience and to which experience needs to conform. He at times used analogies to describe the use of reason as a possession of the spectator rather than an active participant in the acquisition of knowledge.

To illustrate his point, Dewey referred to the legacy of Newtonian laws. Though he did not doubt that Newton employed empiricism to arrive at his discoveries, he contended that since they were posited they have been used as fixed certainties that limit further inquiry rather than extending its bounds. In this regard, Dewey applauded the discovery of Heisenberg's uncertainty principle as introducing openness and flexibility into the process of inquiry. Dewey noted,

The principle of indeterminacy thus presents itself as the final step in the dislodgement of the spectator theory of knowledge. It marks the acknowledgment, within scientific procedure itself, of the fact that knowing is the kind of interaction which goes on within the world. Knowing marks the conversion of undirected changes into changes directed toward an intended conclusion.[7]

To drill down further, Dewey defined intelligence by asserting that

Intelligence ... is associated with *judgment*; that is with selection and arrangement of means to effect consequences and with choice of what we take as our ends. A man is intelligent not in virtue of having reason which grasp first and indemonstrable truths about fixed principles, in order to reason deductively from them to the particulars which they govern, but in virtue of his capacity to estimate the possibilities of a situation and to act in accordance with his estimate. In the large sense of the term, intelligence is as practical as reason is theoretical.[8]

These brief examples summarize much about Dewey's views on the philosophy of science, knowledge, and life in general. He saw these things as dynamic and not static. They are open-ended and fluid. They require active engagement of human inquiry and experientially grappling with empirical and moral problems to arrive at the best responses that emerge out of those circumstances and the process itself. It was an aim of Dewey's philosophy to apply intelligence in order to recognize the good and render what is good secure.

Engagement for Dewey was key. Our experience engages with nature. And for Dewey the "world" was always what we engage with. Intelligence arises from engagement with problems and how those relations are discovered, measured, manipulated, tested, and controlled. Freedom, for Dewey, was not liberation from the world, it was rather a way of acting with the world and in it. As implied, there is no stark division between the acquisition of knowledge in general and scientific inquiry. Nor is there a division between these processes and the molding of character. For Dewey, we were preeminently social beings investing ourselves in solving problems and we become who we are through active interaction with others.

As noted, Dewey saw science and the acquisition of scientific knowledge as a generalized form of knowledge possessed by both the professional specialist and average people as they engage in the processes of living. In briefest terms, Dewey defined the scientific enterprise as "the existence of systematic methods of inquiry, which, when they are brought to bear on a range of facts, enable us to understand them better and to control them more intelligently, less haphazardly and with less routine."[9]

Historian Robert Westbrook identifies five distinct steps in Dewey's path to (scientific) knowledge:

> 1. a felt difficulty or problem ("thinking begins in what may fairly be called a *forked-road* situation"); 2. the location and definition of this difficulty; 3. the suggestion of a possible solution to the problem; 4. reasoned consideration of the bearings of the suggested solution; and 5. further observation and experiment leading to the acceptance (belief) or rejection (disbelief) of the proposed solution. At the heart of this process was the making of inferences, a leap from "what is surely known to something else accepted on its warrant."[10]

It is of the utmost importance to assess these steps not in a formulaic sense but as functions of personality and the product of learned habits. They are habits immediately recognizable to the woman or man of liberal disposition and that are so much under threat from the tribalism and closed-mindedness that looms over our contemporary political and social landscape. For Dewey these habits implied a "willingness to hold belief in suspense, ability to doubt until evidence is obtained; willingness to go where evidence points instead of putting first a personally preferred conclusion; ability to hold ideas in solution and use them as hypotheses to be tested instead of as dogmas to be asserted; and (possibly the most distinctive of all) enjoyment of new fields for inquiry and of new problems."[11] The importance of developing such habits in great measure accounts for the emphasis that Dewey placed on education as a mainstay of his thought.

Dewey's academic career, especially his tenure teaching philosophy at Columbia University (1904–1932), was set within a charged ideological environment against which he worked out his epistemic and political theories. Dewey's commitment to applied intelligence stood in contrast to closed systems such as fascism, Marxism, and doctrinaire religion, which presented the individual with what William James referred to as a "block universe," one in which reality was given in advance of human application or was a product of determinism. Yet Dewey was also aware of the robust appeal of dogmatic ideologies and understood that his theory of knowledge, which did not lead to fixed points or concrete ends, might be found emotionally wanting for many.

In addition to conflicting with culturally bound and politically transmitted ideologies, Dewey recognized that his theory of knowledge, based on active, intelligent inquiry without the security of predetermined answers or solution to problems, came up against deeply ingrained personal proclivities. In an observation that could well be applied to our contemporary situation, Dewey observed in *Freedom and Culture* that the constituents of the scientific attitude are challenged by all too common aspects of human behavior:

> Every one of these traits goes contrary to some human impulse that is naturally strong. Uncertainty is disagreeable to most persons; suspense is so hard to endure that assured expectation of an unfortunate outcome is usually preferred to a long-continued state of doubt. "Wishful thinking" is a comparatively modern phrase, but men upon the whole have usually believed what they wanted to believe, except as very convincing evidence made it impossible. Apart from a scientific attitude, guesses, with persons left to themselves, tend to become opinions, and opinions dogmas. To hold theories and principles in solution, awaiting confirmation, goes contrary to the grain.[12]

A question that can be asked is whether Dewey's theory of knowledge contributes to the relativism we witness today and as such abets the erosion of respect for science and scientific authority. Does Dewey's rejection of a priori truths confirm a culture of subjectivity that assures that "guesses . . . tend to become opinions and opinions dogmas"? In later life Dewey spurned the use of "truth" and opted to employ the term "warranted assertability" to describe the conclusion to which intelligent inquiry led.

The charges of relativism and subjectivism frequently attach themselves to pragmatism, but what saves Dewey from this criticism is his commitment to the social and communal nature of inquiry. New conclusions need to be responsive to the conclusions confirmed by the community of inquirers and new ideas need to cohere with the reservoir of all previously funded ideas. Objectivity is confirmed not by the correspondence of an idea "out there," but by the intersubjective consensus of human beings. Hence Dewey's theory of inquiry assured that there was little wiggle room, so to speak, to irresponsible or wildly assert conclusions emerging out of a method that was designed for an ever fluid reality.

Dewey recognized that a relative few (might we say an *elite* few?) possess the scientific attitude while the masses do not. Yet, it is a disposition that Dewey aimed to spread widely throughout society. Why are the necessary habits of mind so narrowly held? Dewey's response is that "the answer given to this challenge is bound up with the fate of democracy."[13]

DEWEY AND DEMOCRACY

For Dewey, democracy was the process of intelligent inquiry extended to the level of society. The association of democracy with the routine of periodically casting a vote is an expression of democracy in its minimal form. Democratic practice goes beyond the institutions that enable democratic process to take place. At its core, democracy is a character trait of a free people, a trait that dialectically shapes itself through engagement with people and community, and with their problems.

Dewey took for granted the structural requirements of democracy—elections, a free press, an active judiciary, and fundamental rights. But his primary interests did not lie there. Commensurate with the pursuit of knowledge, Dewey was concerned with society shaping itself in order to bring new values to life and to fulfill human desires. In Dewey's own words,

> The keynote of democracy as a way of life may be expressed, it seems to me, as the necessity for the participation of every mature human being in the formation of the values that regulate the living of men together.[14]

But the regulation is not one that maintains static adjustment. Rather, Dewey was a progressive and meliorist, who saw the improvement of the social condition in the capacity of men and women to mold their environments for the better.

It should be clear that Dewey's vision of American democracy was thoroughly secular, and, while optimistic, was totally divorced from any notion of divine purpose or destiny for the nation. Dewey stood opposed to supernaturalism and the authoritarianism that established religion exemplified, all of which was antithetical to the democratic spirit. Democracy is not the working out on the mundane plane a divine assignment or the fulfillment of a religious vision. In accordance with Dewey's overall philosophy, democratic process is itself on the same level as religion as Dewey reframed it, that is, religion of a very humanistic kind.

> His goal was to integrate fully the religious life with the American democratic life, transforming the religious life into a way of practical liberation for the individual and society and the democratic way of life into a way of religious self-realization and social unification.[15]

But none of this relates to blending the worship of a supreme being with the temporal concerns of daily living, or includes the hope that the will of a divine being will be realized in the American experience. As the philosopher Richard Rorty notes, Dewey's democracy "is a matter of forgetting about eternity."[16] In Dewey's words, "Democracy is neither a form of government nor a social expediency but a metaphysic of the relation of man and his experience in nature."[17] Democracy is a thoroughly human process, totally self-contained within human experience without external references points or norms.

Dewey's emphasis on democracy as process suggests that the more important manifestation of democratic participation takes place in the smaller, local spheres of life. No doubt, one needs to be concerned with the issues that consume national and international politics, and assuredly Dewey was. But the heart of democracy is most readily found in what has come to be termed "civil society," that stratum of public situations between the institutions of government and the corporate world. Dewey may have had in mind the New England town meeting as the paradigmatic case. But one can conclude that involvement in improving neighborhood schools, coaching a child's soccer team, or engagement with environmental cleanup campaigns, and finding, in community with others, the best ways to organize these efforts, goes to the heart of Dewey's democratic vision.

Several philosophical concepts intrinsic to Dewey's model are buried within such examples. First is the notion, stated briefly above, that the individual is social and communal in nature. Dewey stood against classic liberalism, which views the individual as separate or independent from others. And certainly Dewey rejected the historical notion that the individual stands in opposition to society so that the gain of one is the loss of the other. A laissez-faire notion of the individual, derived from classical Enlightenment figures, Dewey viewed as the "old liberalism" that needed to be replaced by new values. By contrast, he embraced an organic vision that sees community and active engagement with it as the matrix out of which the individual emerges. This relationship means that Dewey's philosophy is a dynamic one, with social action natural to it.[18]

Second, while Dewey was assuredly a communitarian, he nevertheless was committed to the value of freedom. But in parallel with his philosophy of the social nature of the individual, freedom for Dewey did not mean an absence of restraint or social encumbrance. There is no doubt that Dewey defended individual liberty in a political sense. He played a role in founding the American Civil Liberties Union, and, as such, saw a necessary place for what philosopher Isaiah Berlin would refer to as "negative liberty." Yet in his democratic theory what most concerned him was "positive freedom" in that it was an expression of potentialities that could be realized through involvement in democratic process.

In summary, Dewey was a progressive, a Left liberal with strong affinities for democratic socialism, who believed that struggling for social justice was the beating heart of American identity. He lived and worked in a time of ideological ferment. He was critical of capitalism, yet was thorough in his critique of Marxism for its authoritarianism, determinism, and its emphasis on conflict over cooperation and historical insistence that violence was necessary. As a generalization, Dewey's contribution to democracy was to lay out a theoretical template more than the advocacy of specific policies and programs.

Yet in his long career Dewey spoke out on the issues of the day, defended numerous causes, and joined in association with other progressives. He worked for educational reform and, as mentioned, supported the movement of teachers to unionize and the cause of labor. He reluctantly supported America's entry into the First World War, and he opposed American involvement in the Second World War until after the attack on Pearl Harbor. He was not trusting of Franklin Roosevelt, but he affirmed many of the programs put forth in the New Deal. Dewey lent his voice to, and in several instances chaired, a range of activist organizations. Among them were the League for Independent Political Action and the People's Lobby. These groups brought together coalitions of workers, farmers, and the middle class in order to push for a radical third party. The former also attracted the support of a large group of left-wing intellectuals, including W. E. B. DuBois, Oswald Villard, and Norman Thomas.[19] In addition to the American Civil Liberties Union, he helped found some of the foremost organizations of his day—the NAACP, the League for Industrial Democracy, the New York Teachers' Union, the American Association of University Professors, and the New School for Social Research. Dewey, as noted, set the standard for the public intellectual. The historian Henry Steele Commager observed when Dewey died in 1952, "He was the mentor, and the conscience of the American people; it is scarcely an exaggeration to say that for a generation no issue was clarified until Dewey had spoken."[20]

DEWEY'S RELEVANCE FOR OUR TIMES

The philosopher Richard Rorty assessed John Dewey, together with Walt Whitman, as the foremost exemplars of a defining narrative coursing through American history. It is the narrative of the American Left, but a Left that is detached from any cosmic authority or plan. It is a pragmatic Left, immersed in experimentalism and optimistic in its conviction that, given the requisite intelligence, problems could be solved, new values could emerge out of new circumstances, and human suffering could be ameliorated. It is a Left that is humanistic in its vision and methods. Much of which is most admirable in America's history has been expressed in the struggle for justice and in social reform on behalf of the oppressed and the excluded. John Dewey's philosophy gives peerless intellectual justification to this noble tradition.

Needless to say, America has spun different and darker narratives, those of racism and nativist xenophobia. Imperialism and violence, hegemonic capitalism, social subjection, and marginalization reflect the ignoble underside of American history. At its worst, slavery, genocide, Jim Crow, violent hate groups, and white supremacist organizations are as intrinsic to the American story as the persons and movements that have stood for the better angels of the American experience. And currents of anti-intellectualism have always run through the American character.

The same nation that could elect its first African-American to highest office in short order has replaced him with an unhinged, unread, pathological narcissist and liar, who has unleashed and legitimated the darker expressions of the American national story. It is within this dark context that we can assess John Dewey's relevance.

It is hard to imagine a social condition more antithetical to Dewey's philosophy and politics than the one we experience in 2018. Beyond the attack on science, to the phenomenon of "alternative facts" and "posttruth" assertions, it is hard to know in detail what Dewey would make of the current age. Digital technologies, which have infinitely broadened access to ideas, have also siloed information in ways that intensify opinions and reify them into extremist dogma.

Dewey's theory of science as well as his elaboration of democracy require discussion, mediation, negotiation, and compromise in order to arrive at satisfactory responses, which themselves will be provisional and subject to inevitable change. The contemporary public, from Congress to the grassroots, seems so angrily divided as to foreclose such a process, which at least requires enough openness and respect across lines of difference to allow such a process to take place.

Our times and Dewey's, needless to say, greatly differ, but there are enough similarities in regard to broader phenomena that we can feel confident in concluding where Dewey would stand. He would have decried the hegemonic role of corporatism and the corporate state in suppressing individual freedom.

He would have been aghast at the decline of public education and the imposition of testing as a form of regimentation that directly contradicted the experiential centrality of his educational theory and its lived practice.

In his social theory, Dewey stridently opposed the ideology of the melting pot for robbing minorities, including immigrants, of the richness of their cultural experiences. He was an ethnic pluralist, who nevertheless would have been horrified at the retreat into tribalism that plagues so much of our national life. Ethnic pluralism is only worth having if diverse groups communicate with each other. The point of his philosophy centered around the sharing of different values in order to create richer lives for all concerned.

And Dewey would have railed against the recrudescent nationalism emerging domestically and abroad. Dewey was a cosmopolitan and in principle felt that any type of nationalism or patriotism that blocked the association of people around science, art, and commerce should be abandoned. According to Stephen Rockefeller,

> Dewey continues to adhere to the view he expressed in the early 1890s that the democratic life is a strategy for dissolving all social divisions and for creating world-wide human community. The experience of the two world wars only deepened Dewey's convictions on this matter.[21]

Yet our times may not be completely bleak nor devoid of Deweyan values. The ominous pall of the Trump phenomenon is being met by an outbreak of grassroots activism. In her book *Necessary Trouble*, journalist Sarah Jaffe documents the Occupy Movement, Black Lives Matter, struggles around the country for a $15 minimum wage, efforts to unionize Walmart workers, and other causes, many that fall beneath the radar of the established press, and all carried out by local activists.[22] In what seems like a throwback to an earlier era, public school teachers in West Virginia have successfully concluded a statewide strike to win a five percent pay increase. And hundreds of thousands of high school students have come together to organize protests in Washington, DC, and around the country for sensible gun laws in the wake of a mass shooting at a school in Parkland, Florida. Accurate prediction of the future is not ours, but Dewey, humanist that he was, believed that the future is always open and that the dialectics of change, guided by intelligence, is an antidote to despair.

Robert Westbrook, in his study on Dewey and American democracy, interprets Dewey as more radical than he is generally assessed to have been. He sees Dewey as an influence in the 1960s in inspiring the content of the "Port Huron Statement," which was the manifesto of the Students for a Democratic Society with its call for "participatory democracy." One might conclude as we enter a new age of reaction and rebellion against the forces of repression that Dewey's commitment to participatory democracy may prove to be his contribution of greatest relevance in this moment. We may be at the dawn of a rebirth of democracy that coveys much of the Deweyan spirit.

In the final analysis, the relevance of Dewey's thought today is tagged to the relevance of academia to the everyday realm of human affairs. And a compelling case can be made that what transpires in the ivory tower is less relevant than it was in Dewey's day. But Westbrook argues that, if we seek a blueprint by which to guide our actions and lead us to a better future, "we could do worse than to turn to John Dewey for the measure of the wisdom we will need to work our way out of the wilderness of the present."[23]

CHAPTER 4

THE PHILOSOPHY OF THE OPEN FUTURE

LEE SMOLIN

OPTIMISM: THE WORLD AFTER CLIMATE CHANGE

L et us imagine it is 2080 and the problem of climate change has been solved.[1] Our children will be elderly, or perhaps due to advances in medicine still in the prime of life. How will their thinking have been changed by their avoidance of catastrophe? It is easier to imagine what their perspective will be if we do nothing to bring the CO_2 emissions under control. As they face rising temperatures and sea levels, drought and failing crops, as the northern cities crowd with refugees, one can well enough imagine what they will wish they could say to us.

But suppose human beings find the wisdom to avoid all this. What might we have learned along the way that would make success possible?

It is hard to imagine the consequences of solving the climate crisis because to succeed we have to do more than solve a big engineering problem. Even among those who appreciate the seriousness of the crisis, adherence to one or another of two naive commitments delays real progress. For those who see the world in economic terms, nature is a resource to be exploited and then transcended. For them this is just the problem of agriculture on a larger scale, to be managed by a cost-benefit analysis. Opposite this idea is the romance that underlies much environmental activism, the notion of a pristine nature that can only be diminished by the encroachments of human civilization. For those who hold this belief, the climate crisis is just another issue of preservation. Both sides miss the point, because both assume that nature and technology are mutually exclusive categories, so that when they clash a choice must be made between them.

Any adequate solution to the crisis, however, must muddy the distinction between the natural and the artificial. This is because the climate crisis is not a problem that can be solved with a one-time fix, after which business continues as usual. The problem requires not a choice between nature and technology but a reorientation of their relationship to each other. The overwhelming scientific consensus tells us it is we who are now destabilizing the climate, but it is also true that the climate has in the past fluctuated suddenly between very different states. Were the climate to change dramatically—whether triggered by our doings or not—it would have dire consequences for us. As long as we have the capability to do so, we must prevent or moderate major changes in the climate, for the same reason we must look out for and destroy asteroids that might collide with earth. Once we have resolved the current emergency, we will be committed to a continuing regulation of the climate to keep it in a range where human beings can thrive. This means melding our technologies with the natural cycles and systems that already regulate the climate. By the time we have succeeded in understanding how the systems that regulate the climate react to our technologies, and using this knowledge to further regulate our technologies and economies so that they work in harmony with the climate, we will have transcended the divide between the natural and the artificial on a planetary scale. The economy and the climate will not be two things, they will be aspects of a single system. Thus, to survive the climate crisis we have to conceive of and bring into existence a novel kind of system, which will be a symbiosis of the biological processes that determine the climate with our technological civilization.

The idea of a novel kind of organization merging nature and technology goes against the common presupposition that the natural and the artificial are exclusive categories. We are used to seeing ourselves as apart from nature, and our technologies as impositions on the natural world. But whether we fantasize about transcending nature or about nature surviving us, we have reached the limits of the usefulness of the idea that we are separate from nature. If we want to survive as a species we need a new conception in which we and everything we make and do are as natural as the cycles of carbon and oxygen that we emerged from and in which we participate with every breath. To begin this task, we have to understand the roots of the distinction between the artificial and the natural. As I will argue here, these have a great deal to do with time. The false idea we need to put behind us is what the Brazilian philosopher and politician Roberto Mangabeira Unger has punningly called the perennial philosophy: the idea that what is bound in time is an illusion and what is real is timeless.

Early expressions of the perennial philosophy can be found in the Christian interpretations of the cosmology of Aristotle and Ptolemy, in which the earthly sphere is the unique abode of life but also of death and decay. It is surrounded by perfect spheres of unchanging crystalline construction that rotate eternally around the earth, carrying the moon, sun, and planets. The stars are fixed to the final sphere, above which live God and his angels. From this we get the common notion that goodness and truth are to be found above us, while evil and falseness lie below. To learn to live with our planet we have to grow out of this very old yearning for elevation from it. This hierarchy applies also to the natural/artificial divide, in that the natural is valued over the artificial because it is closer to absolute perfection, and therefore closer to timelessness.

How can we get rid of the conceptual structure of a divided and hier- archical world separating the natural and artificial? To transcend this conceptual trap we have to altogether eliminate the idea that anything is, or can be, timeless, and see everything in nature, including ourselves and our technologies, as time-bound and part of a larger ever-evolving system. There is another way in which the issue of the separation of the natural and the artificial comes down to our conception of time. We human beings have always gotten out of crises by inventing new ideas and novel structures. We have to do this again to survive climate change. But if the search for knowledge is essentially the task of remembering truths that have been ever-present, then the possible solutions to any problem already exist in fixed menus of possibility, and novelty is an illusion. A world without time is a world with fixed categories and fixed possibilities that cannot be transcended.

If, on the other hand, time is real, and everything is subject to it, then there are no fixed categories and no barrier to believing in the act of inventing genuinely novel ideas and solutions to problems. So to understand the distinction between the natural and artificial, and to open up the possibility for novel forms that are both, we have to situate them in time.

To fully realize the conception of a time-bound world, we need to invent a notion of truth that removes timelessness from the equation of truth and replaces it with a notion of truth that is no less objective, but which has room for surprise, invention, and novelty. That is, to have a future, we human beings have to learn to think and talk about the future differently.

A new philosophy is needed that anticipates the merging of the natural and the artificial by achieving a union of the natural and social sciences, in which human agency has a rightful place in nature. This is not a relativism in which anything we want to be true can be. To survive the challenge of climate change, it matters a great deal what is true. We must also reject both the modernist notion that truth and beauty are determined by formal criteria and the postmodern rebellion from that, according to which reality and ethics are merely socially constructed. What is needed is instead a relationalism, according to which the future is restricted by, but not determined from, the present, so that genuine novelty and invention, while rare, are real and possible. This will replace the false hope for transcendence to a timeless, absolute perfection with a genuinely hopeful view of an ever-expanding realm for human action and agency, within a cosmos with an open future.

SCIENCE AND THE FALSE HOPE TO TRANSCEND TIME

The notion that truth is eternal and lies outside of the world we experience is not only a religious idea. It is in science as well, as is exemplified by the Platonic view of mathematics. According to this view, the objects mathematicians study, such as numbers and shapes, exist in an eternal realm apart from physical reality, but are somehow just as real. This view is not popular among philosophers due to a powerful and commonsense objection to it, which is that to the extent that mathematical objects live in a separate realm outside of time and space, there is no way for human beings to gain access to or knowledge about them. Mathematical Platonists, some of them great mathematicians, such as Alain Connes and Roger Penrose, reply that they believe in the power of intuition to give them a transcendent view into this timeless realm. If the Platonists are right, then mathematics is discovery of preexisting truths, so there can be no novelty and no genuine invention. Thus, this view of mathematics reinforces the idea that the scope of human agency and creation is limited to a contents of a preexisting and timeless realm.

The natural world as we experience it differs profoundly from the imagined worlds of theology and mathematics in that, in our world, time is ever-present. We experience the world a moment at a time, each moment one of a succession, and we experience each moment emerging from the previous one, an aspect of time Henri Bergson called *becoming*.

A fundamental question for science is whether or not nature is really like the imagined Platonic realm of mathematics, so that our experience of the flow of present moments, each one a becoming, is an illusion. The claim that our existence in time is an illusion is the heart of the perennial philosophy. There are two versions of this view in physics. In one, called the block universe picture, what is real is actually and only the whole history of the universe, all at once. This was the view of Einstein and many twentieth-century philosophers. (Although, one of them, the logical positivist Rudolf Carnap, recounts conversations in which Einstein regretted the loss of a place for the present moment.) Another version is proposed by the physicist and philosopher Julian Barbour, in his book *The End of Time*: all the possible moments that might be part of the history of the universe exist together in a big heap, so that not only the succession but also the ordering of moments of time is an illusion. If either of these views is right then there is no novelty and no real scope for human agency. The future is no more and no less real now than the present or the past. This timeless view of nature is compatible with the way that physical systems are described in mathematical language in Newton's physics, and equally so in Einstein's general relativity and quantum mechanics. The method whereby time is expelled from the description of a physical system is a codification of the necessity to do experiments over and over again to confirm their results. One can repeat an experiment, varying the initial preparation of the system, and recording the outcome. By doing so one isolates what is general and repeatable in all the instances and calls that a law. The law is then understood to act on the system as it is initially prepared and transform it into its later states. The possible preparations are called the initial conditions. One then takes all the possible states or initial conditions and abstracts them into a timeless "space of states." The law then acts on these states to evolve initial states of an experiment over time to final states. The law is both that which does not change and the generator of change; it gives the unchanging rules for how change takes place.

When applied to describe experiments done in a laboratory this is a very successful method. But it is important to emphasize that it makes sense and accords with experimental practice when applied to a small isolated system, with the experimenter and her clocks and measuring instruments left outside in the larger universe. But if we attempt to apply this picture of a laboratory experiment to the universe as a whole, we commit a metaphysical fallacy. There is no scientific justification for this extrapolation, as there is no experimental result whose explanation requires it. Nonetheless, it is made unthinkingly by most contemporary cosmologists. What helps us fall into this fallacy is the old dream of transcendence, because it requires the scientist to think of himself as outside the universe, watching it without being a part of it. This is then a fantasy of the scientist as god.

One principle of a relational philosophy is that nothing can be outside the universe of events evolving in time, and yet act on it. This implies there is no view of the whole universe as if from outside of it, and no timeless laws that are imposed on the world as if from outside. Since the space of states and laws contemplated in this method are described mathematically, it is easy for a cosmologist to proceed further under the influence of the yearning for transcendence and proclaim that the whole history of the world is identical in every way to an object in the eternal Platonic realm of mathematics. Since both the history of the world and its mirror in mathematics are timeless, this is a vision of nature in which time has been expelled. There is no present moment, and certainly no room for novelty, invention, or human agency.

There is another way that time is expelled from physics. If one considers an isolated system in nature, say the contents of a thermos bottle, one can argue that the system evolves to a state of equilibrium, in which there is no organization or order and no change, apart from the random motion of molecules. This equilibrium state is timeless, and it is unique, so that here too there is no room for surprise or novelty. Some cosmologists go even further, and follow Max Tegmark in identifying the history of the world with the imagined mathematical object to which it is believed to be identical.

This is a case where great metaphysical harm has been done by taking a method that works well when applied to a small isolated system as might be studied in a laboratory and extending it to the universe as a whole. For the universe is not in equilibrium, with its myriads of hot stars pouring photons into cold space. Indeed, life is possible because the earth's surface is continually energized by a flow of the energy carried by those photons through it. Nonetheless, a lot of silly cosmology, from Boltzmann in the late nineteenth century to the fantasies of contemporary quantum cosmologists, imagines that the natural state of the universe is to be in such a lifeless equilibrium, with the presently observed nonequilibrium state only a temporary and rare fluctuation. A favorite model of contemporary theorists is called eternal inflation, whose name signifies that it reflects the old desire for a transcendent science in which the cosmologist stands mentally outside of time and space. This field is as much speculative metaphysics as science. A lot of its literature is taken up with issues that can have no bearing on experiment, such as how probabilities are to be defined in an infinite and timeless realm, or whether it is overwhelmingly probable that a conscious brain would appear as a fluctuation in an eternal state of equilibrium rather than as a product of biological evolution.

Part of the program of the new philosophy is to save cosmology from this unscientific excursion by recognizing the central role time must play on a cosmological scale. But more importantly, a civilization whose scientists and philosophers teach that time is an illusion and the future is fixed is not likely to be able to summon the imaginative power needed to invent the communion of political organizations, technology, and natural processes needed for human beings to thrive sustainably beyond this century.

THE TROUBLE WITH ECONOMICS

Probably the greatest harm done by the metaphysical view that reality is timeless is through its influence on twentieth-century economics. The basic flaw in the thinking of many economists is that a market is a system with a single equilibrium state. This is a state where the prices have adjusted so that the supply of each good exactly meets the demand for it, according to the law of supply and demand.[2] Furthermore, it can be argued that the equilibrium point of a market optimizes everyone's satisfaction. There is a mathematical theorem that in equilibrium no one can be made happier without making someone else less happy.

If each market has one and only one such equilibrium point, then the wise and ethical thing to do is to let the market alone so it can adjust to that point. Market forces—i.e., the way producers and consumers respond to changes in prices—should be sufficient to do this. A recent version of this idea is the efficient market hypothesis, which holds that the prices reflect all the information relevant to the market. In a market with many players contributing their knowledge and views by means of their bids and asks, it is impossible, in principle, that any asset is for very long mispriced.

It is remarkable that this line of reasoning can be backed up by elegant mathematical proofs, in mathematical models of market economies. Within these models, there are even formal proofs that equilibrium points always exist, there are always choices of prices such that supply exactly balances demand.

However, this simple picture in which the market always acts to restore conditions to equilibrium depends on the assumption that there is only one equilibrium. But this is not the case. Economists have known since the 1970s that their mathematical models of markets have typically many equilibrium points where supply balances demand. How many? The number is hard to estimate but certainly grows at least proportional to the numbers of companies and consumers, if not faster.

In a complex modern economy with many goods, each made by multiple firms, and bought by a large number of consumers, there are many ways to set the prices of all the goods so that supply and demand are in balance. This is shown by a theorem proved in 1972 by three highly influential economists. One of them was Hugo Sonnenschein, who was not just a member of the Chicago school but served as president of that university.

Because there are many equilibria where market forces balance, they cannot all be completely stable. The question is then how a society chooses which equilibrium to be in. It cannot be by market forces, because supply and demand are balanced in each of the many possible equilibrium. There is then a necessity for other forces to determine which way the market forces are satisfied. Hence there is a necessary role for regulation, law, culture, and politics to play in determining the evolution of a market economy.

How is it possible that influential economists have been arguing for decades from the premise of a single, unique equilibrium, when there were results in their own literature by prominent colleagues that showed this was incorrect? I believe the reason is the pull of the timeless over the time-bound. For if there is only a single stable equilibrium, the dynamics by which the market evolves over time is not of much interest. Whatever happens, the market will find the equilibrium, and if it is perturbed it will oscillate around that equilibrium and settle back down into it. You don't need to know anything else.

If there is a unique stable equilibrium, then there is not much scope for human agency (apart from each firm maximizing profits and each consumer maximizing their pleasure), and the best thing to do is to leave the market alone to come to equilibrium. But if there are many possible equilibria, and none is completely stable, then human agency has a big role to play, to participate in and steer the dynamics by which one equilibrium is chosen out of many possibilities. What seems to have happened, in the thinking of the economic gurus who won the day for deregulation, is that the role of human agency was neglected in deference to an imagined mythical timeless state of nature. This was the profound conceptual mistake that opened the way for the errors of policy that led to the recent economic crisis and recession.

Another way to speak about this mistake is in terms of path dependence and independence. A system is path dependent if there are many ways it could evolve, so that some measure of choice is involved in where it goes. A system is path independent if there is no choice and there is only a single state to which the system can evolve. In a path independent system time and dynamics play little role, because at any time the system is either in its unique state or at worst, fluctuating slightly around it. In a path dependent system time plays a big role.

Neoclassical economics conceptualizes economics as path independent. An efficient market is path independent, as is a market with a single stable equilibrium.

In a path independent system it should be impossible to make money purely by trading, without making anything. Opportunities to do so are called arbitrage, and there is a basic understanding in financial theory that in an efficient market arbitrage is impossible because everything is already priced in such a way that there are no inconsistencies. You cannot trade dollars for yen, trade those for Euros, and trade those back for dollars and make a profit. Nonetheless, hedge funds and investment banks have made fortunes from trading in currency markets. Their success should be impossible in an efficient market, but this does not seem to have been a problem for economic theorizing.

Decades ago, Brian Arthur, who was at the time the youngest holder of an endowed chair at Stanford University, began to argue that economics was path dependent. His evidence for this was that a law of economics, called the law of decreasing returns, is not always correct. Decreasing returns is the idea that the more of something you make, the less profit you make from each item you sell. This is not necessarily true, for example, in the software business. Arthur's work was treated as if it were heretical, and indeed, without the assumption of decreasing returns, some of the mathematical proofs in neoclassical economic models cannot be carried through. He ended up leaving Stanford and founding the economics program at the Santa Fe Institute.

In the mid-1990s, an economics graduate student at Harvard, Pia Malaney, working with a mathematician, Eric Weinstein, found a mathematical representation of the path dependence of economics. In geometry and physics, there is a well-understood technique for studying path dependent systems, which is called gauge fields. They provide the mathematical foundations for our understanding of all the forces in nature.

Malaney and Weinstein applied this method to economics and found that they are path dependent. Indeed, there is a quantity that is easy to compute, called the curvature, that measures path dependence, and they found it was not zero in typical models of markets where prices and preferences of consumers are both changing—i.e., in the real world.[3]

The work of Malaney and Weinstein was ignored by academic economists, but the basic fact that markets are path dependent has been since rediscovered by a number of physicists who took their skills to the financial markets and found it natural to apply gauge theories to the markets. Hence, like the earth, and the geometry of space-time, the mathematical spaces that model markets are not flat.

There is no way of knowing how many hedge funds are making money by discovering arbitrage opportunities by measuring curvature—i.e., path dependence that is not supposed to exist in neoclassical economics—but it is hard to imagine this is not going on.

A path dependent market is one where time really matters. How does neoclassical economic theory deal with the fact that in reality markets do evolve in time, in response to changing technologies and preferences, continually opening up opportunities to make money that are not supposed to exist in their models? Neoclassical economics treats time by abstracting it away. In a neoclassical model you as a consumer are modeled by a utility function. This is a mathematical function that gives a number to every possible combination of goods and services that might be purchased in the economy you live in.

This is a rather huge set, but hey, this is math, so let's continue. The idea is that the higher the utility a collection of goods and services has for you, the more you would like to buy them. The models then assume that you buy the collection of goods and services that maximizes your desires, as measured by your utility function, given the constraint of how much you can afford.

What about time? The idea is that the lists of goods and services includes all the goods and services you might want to buy over your entire life. So the budget constraint that is imposed is over your lifetime income. Now this is clearly absurd—how could anyone have any idea what they will want or need decades from now, or what one's lifetime income will be?

What about contingency (i.e., the fact that over our lives we will confront a myriad of circumstances that cannot be predicted)? The models deal with these by lumping them into the lists of goods and services (i.e., the assumption is that there is a price for every possible collection of goods and services at every time and in every contingent situation that might arise. There is a price not just for a Chevy Mustang, but for a Chevy Mustang in 2020 under every possible contingency). The models assume not just that all the goods we might buy now have been perfectly priced in equilibrium, but that every future price of any collection of goods under every possible contingency is also perfectly priced. This is still more absurd, yet this is a description of the theory used and taught by the economists who advise governments.

The fact that the neoclassical economic models go to such absurd lengths to abstract time and contingency away shows how central the issue of time is. There is a powerful if unconscious attraction to theories in which time plays no role. It gives theorists the impression of standing outside the world, in a timeless realm of pure truth, against which the time and contingency of the real world pale.

It took a string theory postdoc at our theoretical physics institute, Samuel Vázquez, less than six months to go from hearing these ideas to measuring path dependence in real market data.[4] What he was doing was impossible and heretical in the framework of neoclassical economics theory, but there it was in real data.

We live in a world in which it is impossible in principle to anticipate and list most of the contingencies that will arise in the future. Neither the political context, nor the inventions, nor the fashions, nor the weather nor climate are specifiable in advance.

There is in the real world no possibility of working with an abstract space of all the contingencies that may evolve in the future. To do real economics, without mythological elements, we need a theoretical framework in which time is real and the future is not specifiable or knowable in advance, even in principle. It is only in such a theoretical context that the full scope of our power to construct our future can make sense.

Furthermore, to meld an economy and an ecology, we need to conceive of them in common terms. The grounds for a common conception are in seeing each of them as open complex systems, evolving in time, with path dependence and multiple equilibria, governed by feedback. This fits the description we have briefly given here of economics, and it fits the theoretical framework of ecology, with the climate as the sum and expression of a network of chemical reactions driving and regulated by the basic cycles of the biosphere.

SCIENCE AND THE OPENNESS OF THE FUTURE

If our civilization is to thrive, it would be very helpful to base our decision-making on a coherent view of the world, in which there is a union between natural and social sciences. If we are to purge from economics and social theory the notion that change is an illusion hiding timeless truths, we should do this in science as well. And we should do this all the way down, to the foundations of physics, cosmology, and mathematics.

One place to start is with the limitations on the notion that everything can be reduced to laws of physics. Reductionism is the good advice that if you want to understand the workings of a complex system, it is good to have a working knowledge of the properties of the parts. But this does not mean that the properties of the whole can always be completely reduced to sensible and useful statements made entirely in the language used to describe the parts. Biological cells are made of molecules, for sure. And it is the case that the properties of molecules can be entirely described in terms of quantum states, which provide probabilities for each set of positions that their component atoms might take. However, cells have properties that can be understood only functionally, such as by their contribution to their fitness. The fitness of a biological cell certainly depends on the molecules that make it up, but this does not mean that there is any useful definition of that fitness as a function of the positions of the atoms that comprise those molecules. Neither in practice nor in principle need there be a complete reduction of the properties needed to explain the higher-level system in terms only of the properties used to describe its component parts.

This means that new concepts, properties, and laws may emerge at a higher level that are not able to be stated in the language used to give a complete description of its parts. This means that novel laws may emerge in time, when the complex systems they apply to evolve into existence.

Four billion years ago there was nothing on earth that the laws of Mendelian genetics or sexual selection applied to. Those laws came into existence only when the cells they apply to evolved.

Moreover, the higher-level laws can, and often do, influence the lower levels of description. There are molecules that only exist on earth because they are components of biological systems and hence byproducts of evolution. Suppose we wanted to know how many hemoglobin molecules are present on earth. Hemoglobin is just a big molecule, completely describable by the laws of quantum mechanics. But there is no way the question of how many exist at a given time could be arrived at just by solving the laws of quantum mechanics. To explain why there are any hemoglobin molecules on earth, and to estimate their number, one must reason in terms of higher-order laws that apply to animals and are not completely reducible to quantum mechanics. For one thing, to determine whether evolution would lead to the emergence of animals, and hence blood, one must reason in terms of selective advantage.

Applying this to the future, it is fair to say that one cannot, even in principle, propose any calculation to give a list of the molecules that will be present in the biosphere in a billion years. Evolution sometimes proceeds by discontinuities, by which novel forms of organization emerge. These have included cells, eukaryotic cells, multicellular life, plants, animals, intelligence, language, etc. Each of these was an unprecedented novelty when it emerged. There is no reason to believe that evolution has produced its last novel form of organization, and no reason to believe that any scientific method could be devised that would provide a complete list of such possible novelties. Surprise is a real part of science, even if the laws of physics that apply to elementary particles are deterministic.

How does this ubiquity of novelty square with the notion that time is absent in physics?

As we have seen, in physics as developed so far, there are timeless laws acting on timeless spaces of possible states. On the level of complex systems there are neither.

The space of states in the biosphere or an economy can suddenly expand, in ways that are not predictable from the present state. As they do, novel forms of organization can emerge, which can exhibit novel regularities described by novel laws. The best that theory may be able to do is describe what complexity theorist Stuart Kauffman calls the adjacent possible—that is, to discern some of the very next forms that may emerge. Or there may be singularities where not even this is possible.

Another limit to the applicability of reductionism comes at the lower end. If the properties of the whole depend on the parts, what determines the properties of the parts? One can answer, break it down further, into smaller parts. This has proceeded well, from molecules to atoms to nuclei to quarks, but at some point, if there is to be an end to the chain of explanation, we must get to a particle that is truly elementary, that has no parts. It will still have some properties, which govern how it moves and interacts with other truly elementary particles. What will determine those properties?

This is a part of the question of what chooses the laws of physics. Till recently, this was not a question physicists often asked; our job was to discover what the laws of physics are. But over the last three decades we have been asking more and more, why these laws and not others? Why this set of elementary particles, with the particular properties they have, and not others with other properties?

Many theorists used to imagine that the requirement that the laws of nature be expressible in mathematics would limit a unified theory to a unique candidate. It was felt that the requirements that the laws be mathematically consistent and unified would together only have one solution. So far as we can tell, this is false. The most developed candidate for a unified theory so far, string theory, appears to come in an infinite number of versions, all equally consistent. Motivated by an analogy with landscapes of possible genetic sequences, this is called the landscape of possible string theories. So the mystery of why these laws and not others has been deepened by the discovery that there appear to be many possible consistent laws.

The first person to ask this question cogently seems to have been the founder of the American pragmatist school of philosophy, Charles Sanders Pierce. He wrote, in 1891, "The only possible way of accounting for the laws of nature and for uniformity in general is to presume them results of evolution."[5] In fact, it is possible to invent a scenario by means of which a process analogous to natural selection has acted in our past to produce our universe. It offers a partial answer to the question of why these laws, and it does so in a way that makes genuine predictions that are vulnerable to being shown false by experiment.

This provides further motivation for believing in the reality of time and the openness of the future. If the selection of the laws of nature that determine what elementary particles exist and how they interact is a result of evolution analogous to natural selection, then there must be a time that this evolution plays out in. And it must be a time that exists prior to the laws, for the laws must have scope to evolve in time. Moreover, if the process of natural selection can lead to novel forms of organization in the biosphere, there is no reason to think they cannot do so in the whole of nature. The big bang may even be an event that brought into being a novel form of organization in which particles move in space; if this is so then the laws we probe with particle accelerators like the LHC (Large Hadron Collider) emerged then.

To have a cosmology in which time is more basic than laws, we need an alternative to the Newtonian way of doing physics based on timeless laws acting within timeless spaces of possible states. This is possible because the context in which that method works requires modeling a small part of the universe, with the clocks and observers removed.

There is no empirical result that calls for the science of cosmology to be based on the extension of this method to the universe as a whole. Indeed, attempts to do so result in the endless chasing of absurd questions with no bearing on observation, such as what determines the laws and what chooses the initial conditions. We must find instead a way of framing principles and hypotheses about the cosmos as a whole that does not at the beginning separate the truths of the world into timeless laws and initial conditions.

The starting point for any cosmological theory must be the relational conception of time and space as formulated by Leibniz, which is fully realized in physics within Einstein's general theory of relativity. Within a relational theory, all properties of an entity refer to relationships it has with other entities. Quantum mechanics also partly realizes a relational conception of properties. The revolution started by Einstein in 1905 remains to be completed by a full unification of quantum theory with relativity, and the key issue here is the role of time. Much research in quantum gravity and cosmology has been based on a timeless picture in which time is to emerge only when the universe is sufficiently big. There is a growing sense that these attempts to construct a consistent unification of gravity with quantum theory on a timeless framework fail on their own terms. At the same time, a lot of progress has recently been made on approaches to quantum cosmology in which time is fundamental rather than emergent. There remains a great deal to do in these directions, so what can be said now is only that the idea that time is real rather than emergent is now motivating some of the most interesting research on the cutting edge of the field of quantum gravity.

DEMOCRACY AND THE OPEN FUTURE

One reason to be optimistic about a consilience of the social and physical sciences is that in the past great conceptual steps in one have been echoed in the other. Newton's idea of absolute time and space is said to have greatly influenced the political theory of his contemporary John Locke. The notion that the positions of particles were defined with respect, not to each other, but to absolute space, was mirrored in the notion of rights, defined for each citizen with respect to an unchanging absolute background of principles of justice. As general relativity moves physics toward a relational theory of space and time in which all properties are defined in terms of relationships, and in which the future is open, is this mirrored in an analogous movement in social theory? I believe that it is and that it can be found already in the writings of Unger and a number of other social theorists.[6] These can be understood as exploring in the context of social theory the implications of a relational philosophy, according to which all properties ascribed to actors and agents in a social system arise from their relationships and interactions with each other. As in a Leibnizian cosmology, there are no external timeless categories or laws. The future is open because there is no end to the novel modes of organization that may be invented by a society as it continually confronts unprecedented problems and opportunities.

This new social theory has the task of moving democracy toward a global form of political organization, adequate to guide the evolution of the multi-ethnic and multicultural societies now in ascendance. It must also be adequate to make the decisions necessary to survive the crisis of climate change.

Here I would like to very briefly sketch my understanding of what democracy looks like from the relational perspective of the new philosophy. Remarkably, the same ideas provide an understanding of how science works. This is important, as the challenge of climate change concerns the interaction of science and politics. Both democratic governance and the workings of the scientific community have evolved to manage several basic facts about human beings: We are very smart, but we are flawed in characteristic ways. We have the capacity to study our situation in nature over a single lifetime, and with culture accumulate knowledge over many lifetimes. But we also have evolved the capacity to think quickly and act decisively at the snap of a twig.

This means we often make mistakes, and we often fool ourselves. To combat our tendency for error, we have evolved societies that embrace the contradiction between the conservative and the rebel, in the service of future generations. The future is genuinely unknowable, but the one thing we know for sure is that in the future our descendants will know a lot more than we do. By working within communities and societies, we can achieve much more than we can as individuals, and yet progress requires individuals to take great risks to invent and test new ideas and viewpoints.

Both scientific communities and the larger democratic societies they evolved from progress because their work is governed by two basic principles. Both are necessary for knowledge to develop and wise decisions to be made in the face of a world where the present is increasingly comprehensible, but the future remains unknowable.

1) When rational argument from public evidence[7] suffices to decide a question, it must be considered to be so decided.

2) When rational argument from public evidence does not suffice to decide a question, the community must encourage and nurture a diverse range of viewpoints and hypotheses consistent with a good faith attempt to develop convincing public evidence.

That is, we respect the power of reasoning when it is decisive and, when it is not, we respect those who in good faith disagree with us. The limitation to people of good faith means those within the community who accept these principles. Within such communities, knowledge can progress and we can strive to make wise decisions about a future that is not completely knowable.

These are ethical principles, and communities that adhere to them may be called ethical communities. Such communities are formed and bonded together by a shared adherence to a set of ethical principles. Membership has nothing to do with history, ethnicity, gender, race, or any other circumstance, other than adherence to a set of ethical principles.

There is, of course, a lot of incompleteness in the adherence to these ethics, in scientific communities and democratic societies alike. There is a lot of corruption and taking advantage of power, wealth, and status in both contexts. But what matters in the long run is not perfect adherence, it is the shape and direction of things, and when push comes to shove these are the ethics that make scientific communities, courts and parliaments function.

Scientific communities and democratic societies share a frank acknowledgement that they do not know the future. They educate not to indoctrinate but to prepare future generations to face an uncertain future in which the only thing we can ensure is that they will know more than we do now. We can call these open communities.

There are also ethical communities that are not open, because their ethics is based on a confidence that they do know the future. These include fundamentalist religious communities and some communities of political ideologues. Instead of being open, they have a theory of the future that they believe gives them the right to power over others.

Fundamentalist communities and open communities can be distinguished by their understanding of time and the possibilities for knowledge of the future. Fundamentalists believe that the future already exists, and they are prone to believe that time is altogether an illusion. They believe in mythological stories according to which the ever-changing world we perceive is an illusion that hides a true timeless reality. They teach faith and acceptance and fear of what the future may bring. Open communities teach that the future is not yet real, what is real instead is time and the processes by which the future unfolds and emerges out of the present. For a fundamentalist, human agency is an illusion; for a member of an open society human agency is the necessary means of constructing a future that would not otherwise exist.

People of many histories, religions, and ethnicities can participate in a scientific community or a democratic society, and there is something essentially universal and universalizing about them. Any two people of good faith, who accept the principles stated above, can work together to understand the present and build a common future.

The scientific community certainly demonstrates this. While its origins may have been localized, by now it is completely universal. One sees people from every continent, culture, and religion working together in laboratories and institutes around the world. The same is to be said for the governance of more and more of our democratic states, as immigration becomes more and more central. Canada, the United States, and some European countries have discovered that democracy thrives in communities of people from diverse backgrounds. There is a genuine sense in which there is a common society formed by the democratic nations of the world.

To successfully tame climate change, and then over a century build a sustainable marriage of nature and technology that can keep the climate in balance, will require an unprecedented level of cooperation among the nations, peoples, and corporations of the earth. To make this unprecedented international cooperation possible, there must grow up a recognition of a global culture and community. This is not a replacement for the many cultures of the planet, which will no doubt continue to thrive.

Many of us have learned to live with multiple identities and affiliations, and this will be one more; the romantic concept of a citizen of the planet will gain concrete reality as decisions are made by a global process in the interests of our thriving on this planet.

The basis of this global community will be science and democracy, as captured by the two ethical principles I proposed above.

Over the next few decades, each nation, each community, and each individual will be faced with a choice. They can join this global community by accepting the principles of science and democracy and participating in the decision-making over melding our human technologies with the natural processes of the planet. Or they can choose to stay outside of this universal community. We cannot expect that everyone will join, but we can hope that enough will so that we can make the decisions necessary for our survival.

A NEW VIEW OF ECONOMICS AND ECOLOGY UNFOLDING IN TIME

From the perspective of the new philosophy, the problems that democracy and science must solve together take on fresh perspectives. As we have seen, the problem of inventing and regulating a system of economic markets looks completely different from this perspective. The basic question is no longer to let the market alone to come to a mythical, unique state of equilibrium. It is instead to give full scope to human agency to design a system in which ethical concerns guide the choice of which among many paths the markets evolve along, and so choose, always temporarily, among the many possible states of equilibrium. Most centrally, the menu of possibilities is not fixed, but ever increasing, because of the human capacity for inventing novel products and processes and entire forms of economic organization. As Unger proclaims, the role of intellectuals and politicians is less to debate existing possibilities and more to create novel ones.

This view of economics fits into the larger picture of the biosphere as a self-organized system, far from equilibrium, driven to ever-higher levels of order by the flow of energy from the sun. This view sees the planet as an evolving network of processes, stabilized by feedback processes occurring over a wide range of time scales, from days to eons.

These processes include the carbon, nitrogen, oxygen, and other cycles that are the basis for life and the climate.

The economy is then fundamentally a novel form of organization we human beings invented to participate in and extend those natural cycles and systems. This is illustrated by the fact that the main fuels our economy runs on—so far—exploit buried deposits of carbon. Our food comes from exploitation and extension of those natural cycles, as do most of our medicines. Agriculture and biotechnology are just steps along the path by which we humans have integrated ourselves in the natural cycles and processes of the biosphere.

This perspective completely alters how we think about climate change. There is no single natural state for the earth's climate, that human beings can either leave alone or damage. The climate is not static, and it is not to be taken for granted; it is an aspect of the living planet, determined by processes that involve the gases that cycle among the living plants, animals, and microbes. The fact is that the earth's climate has spent most of the time since life arose in states different from the present. For much of the time, the climate was much warmer than it is now; during the ice ages it was much colder. The record shows that it has sometimes changed abruptly from one state to another and that there is a strong correlation between elevated CO_2 levels and elevated temperature. The science suggests that what stabilizes the climate, when it is stable, is feedback working through great cycles, such as the one by which carbon is transmuted through the atmosphere and oceans from rock to plants to animals and back again.

The present crisis is brought about because our input of carbon and other greenhouse gases into these cycles is on a track to more than double the amount of CO_2 in the air.

There is a lot more to know about the detailed observations and modeling that strongly support the claims that the temperature has risen because of our input of carbon, and will continue to do so long as we continue to pump carbon into the air, and for at least a century afterward. But it is not necessary to know all the details to understand that this is a fundamental crisis whose solution will require long-term changes in how our economies interact with the natural cycles of the planet to produce our food and energy.

The view of nature as pristine and in need of protection has played an inspirational role during the history of environmental movements. Nonetheless, it is based on an antiquated nineteenth-century view of nature that is based on false oppositions between human life and the rest of life, the time-doomed versus the timeless, and the artificial versus the natural. This view is now getting in the way because it gives us no options to resolve the climate-change problem morally, that is to say without giving up the rightful desires of the world's people for freedom from poverty and want. As Stewart Brand has documented in his eloquent book *Whole Earth Discipline*, the romantic idea that equates wisdom with protection of a pristine nature from civilization has led some environmentalists to make the wrong call on a range of issues including nuclear energy and urbanization.[8]

The main mistake is to suppose that there is a unique state for nature and a unique natural path for the evolution of the biosphere. Since there are many possible partly stable states the climate can be in, and many possible paths along which it can evolve, and since the same is true for the economy, we must become part of the processes that choose those states and paths. This means we must join our civilization and technology and the natural systems of the biosphere into a single, novel form of organization.

Since the basic processes on which our economies are built are just extensions of the basic cycles of the biosphere, we are already well on the way to a joining of the natural and artificial. What is required is to consciously accept our role as partners as we modify our systems to bring them into harmony with those of the planet. This means among other things that we cannot simply impose an invented vision of the climate on the planet. We must work cautiously and deliberately, step-by-step, testing and refining our inventions and strategies as we see how the natural systems respond to them.

There is real danger in getting this wrong, but we have no choice, as there is even greater danger in doing nothing.

When it comes to policy, this means that we do not have the luxury of seeing climate change as an emergency, to be dealt with by making some decisions now, after which everything will work out and we can move on to other things. It has always been inevitable that sooner or later the industries of food and energy we need to thrive, being extensions of natural processes, would grow to a scale that their management affected the stability of those processes. What we have to do is learn how to manage our role in, and contribution to, the climate. Like farming, this is not something we do once, this is a responsibility we will have from now on. So we have to understand the next several decades as the first step of a permanent expansion of our role in nature, as consequential as the invention of agriculture. It will take some time to get the hang of it.

While we are doing so, the best policy will be to not try to predict one outcome or choose one path but to expand the range of options available to policymakers in each decade of our century, so that whatever arises there will always be tools available appropriate to the situation then.

So while we certainly have to begin now to drastically lower our contributions to the CO_2 levels of the atmosphere, by all means at our disposal, while developing and putting to use alternative means of energy production, we have to simultaneously work on improving the science of the biosphere to the point where we reliably understand the systems we are interfering with and how to affect them. We have to be prepared to respond to various instabilities and nonlinearities that may drive the climate away from states comfortable for us. At the same time, we have to work on developing modes of international governance adequate to make the decisions that will be required to keep the climate stable and hospitable.

The more tardy we are about implementing aggressive strategies to lower our CO_2 emissions, the more likely it is we will have to implement emergency measures for rescuing vulnerable populations from the consequences of global warming and forestalling the effects through geoengineering. Each of these levels of response must be coordinated and global. Thus, while it is hard to predict how the crisis will evolve, there is no good outcome that does not involve very high levels of international cooperation as well as unprecedented coordination of scientists and policymakers.

There is no choice about whether measures that interdependently regulate the economy and the climate are put into place, the only option is the time scale over which the new governance and regulatory structures necessary to solve the problem are invented. The earlier we start, the more gradually will we be able to implement the changes, the less imposition there will be on economies and the sovereignty of nations, and the sooner the benefits will outweigh the costs.

By the end of this century we will either have failed or there will be a new entity, satisfying new regularities: a planetary civilization that has unified its biology and technology. By doing so we will have done much more than save human civilization from catastrophe, we will have gained the knowledge to fully cultivate our planet as a garden for ourselves and all the living creatures. This may well will involve leaving much of it wild; what must be domesticated is our technologies and economies, together with the feedback processes that stabilize the global climate. And whatever the costs along the way, in the long term there will be enormous benefits to having achieved a global coordination of economies and ecologies necessary to live sustainably on this planet. If the nations of the earth can learn to work together to solve this problem, other key global challenges will also be easier to solve.

As living, intelligent creatures who are products of evolution within this world, we have the blessing of being conscious, active agents, able to invent and construct novel solutions to the surprising situations we are continually presented with. Let us use the full scope of this power and agency to make a world where we and everything else that lives on this planet can thrive sustainably. Let us make a future for our descendants that will bequeath to each generation the freedom and range of choices they need to continually deepen our humanity, while melding our created worlds with nature, so that each generation will make a world that their ancestors could not have imagined, but would nonetheless have thrilled to. Let us, at the very least, not screw it up.

PART II

SCIENCE'S DEMOCRATIC DIMENSIONS

CHAPTER 5

THE SCIENTIFIC REVOLUTION AND INDIVIDUAL INQUIRY

DIANA M. JUDD

Science and democracy do not mix.

A republic, on the other hand, is a different matter. Democracy is a term that gets thrown around far too easily, as if it were some kind of political vocabulary quiz, wherein the object of the game is to define a term as vaguely as possible and hope everyone will approve and no one will question it. In this chapter, I will argue that it is a republic, not a democracy, that contains three core ideas and ideals that we all seem to take for granted: individual liberty, individual rights, and the rule of law. These ideas and ideals stemmed directly from the development of modern natural science. Democracy, in the direct translation of that word from the Greek, "rule by the people," contains none of these elements. A republic *does*. Crucially, it was the development of the modern scientific method that, above all else, brought the intellectual and social tools to bear in the creation of these political ideas.

Even Aristotle knew that democracy was a "deviation" from a "correct" form government. In *The Politics*, he stated that "it is proper to consider not only what is the best constitution but also what is the one possible of achievement, and likewise also what is the one that is easier and more generally shared by all states [and] could be preserved for the longest time."[1] In his view, there were only three "correct" types of regimes: a monarchy, which in the original Greek meant "rule by the one"; an aristocracy, which meant "rule by the best"; and what he termed a "polity," which he described as a form of government wherein the people had an indirect say through the election of representatives in how laws should be formed. He declared that the deviate forms of these three were tyranny, rule by one dictator; oligarchy, rule by the few; and democracy, wherein the populous makes decisions collectively on every law of governance.

In Book Four of *The Politics*, he carefully describes the different forms democracy can take, and the potentially disastrous pitfalls of each. His conclusion? When the people themselves rule directly, there will be self-interests that will inevitably result in injustice to much of the population. For when the population votes, the majority always wins the day. Thus, fifty-percent plus one person of the population will always make the final decision for all the rest. Thus, he concluded, "for where the laws do not rule there is no regime. The law should rule in all matters, while the offices and the regime should judge in particular cases."[2] The form of government that fulfills these criteria is what Aristotle called a polity. We today call that a republic.

The framers of the Constitution of the United States agreed. As James Madison so eloquently pointed out in *The Federalist Papers*, democracy can in no way ameliorate, let alone cure what he called "the mischiefs of faction." Democracies, he argued, "have ever been spectacles of turbulence and contention" wherein there is "nothing to check the inducements to sacrifice the weaker party or an obnoxious individual."[3] A republic, on the other hand, which in his words is "a form of government in which a scheme of representation takes place, opens a different prospect and promises the cure [for faction]."[4]

The framers of the Constitution in general, and James Madison in particular, knew full well that a just government was one that was administered by and for the people. He also knew full well that people are not perfect. According to him, if there is any hope that justice is to be maintained, the powers of government must be separated, and a system of checks and balances must be in place. Relying on one leader is not the answer. As he put it, "It is vain to say that Enlightened statesmen will be able to adjust these clashing interests and render them all subservient to the public good. Enlightened statesmen will not always be at the helm."[5] Instead, he proposed a different solution to this thorny political problem:

Ambition must be made to counteract ambition. The interest of the man must be connected to the constitutional rights of the place. It may be a reflection on human nature that such devices should be necessary to control the abuses of government. But what is government itself but the greatest of all reflections on human nature? If men were angels, no government would be necessary. If angels were to govern men, neither external nor internal controls on government would be necessary. In framing a government which is to be administered by men over men, the great difficulty lies in this: you must first enable the government to control the governed; and in the next place oblige it to control itself.[6]

Thus, while he does acknowledge that the primary control on government must be the people themselves, he is careful to conclude that "experience has taught mankind the necessity of auxiliary precautions."[7] In other words, we need a plan B. That plan is a republic. It is characterized by the separation of powers within the government, and most importantly, by the rule of law. Laws no one is exempt from, not the rulers nor the ruled, and laws that are public, transparent, and applied universally.

Therefore, to ask if natural science and democracy are in any way compatible, the answer is a resounding No. But to ask if natural science and a republic are compatible, the answer is an emphatic Yes. In fact, historically, philosophically, and ideologically, they are *inextricably linked*.

Natural science contains an ethic. I will elaborate.

The key to the ethic of natural science unlocks a door to nothing less than the origins of how an entirely new form of political thought came into being. A form of political thought based on the rule of law, which led to the ideas of individual liberty, individual freedom, and the audacious notion that we are capable of thinking for ourselves. To understand this remarkable progression? We are going to need some Bacon.

In 1620, which I date as the true dawn of the Enlightenment, Francis Bacon wrote the most influential of his works, *The New Organon*. "Organon" in Latin means tool, hence he intended to lay out a completely new method, a new way of thinking about the natural world. In the process, he did nothing less than inspire the inception of modern natural science, which in turn led to a revolution in political thought.

Bacon had published a work in 1605, *The Advancement of Learning*, which was his earliest attempt to formulate his thoughts on the status of knowledge and learning and the insufficiency of both as practiced. The work contains two parts. In the first, Bacon lauds the excellence of knowledge in general and learning in particular, and the value of augmenting and propagating both. In the second, he takes stock of which subjects have been embraced and which have been undervalued. Bacon begins with an argument as to why learning itself must be advanced, and why it is against neither the will of God nor the interests of the state to do so. Bacon wished to deliver knowledge "from the discredits and disgraces which it hath received, all from ignorance, but ignorance severally disguised; appearing sometimes in the zeal and jealousy of divines, sometimes in the severity and arrogancy of politicians, and sometimes in the errors and imperfections of learned men themselves."[8] Thus, he states that the clergy considers knowledge as something to accept with great caution and only within limits. As he put it, for them to desire too much knowledge would be tantamount to giving in to temptation, and therefore sinful. Bacon's argument against this well-known objection was made on religious grounds, while he was careful to emphasize that knowledge should not be sought out of vanity.

It is crucial to remember that before Francis Bacon wrote, it was taken for granted in the West that all "legitimate" forms of knowledge had long been established, thus philosophically and religiously authoritative. In short, they were held to be unquestionable. Beginning a theme that would dominate *The New Organon*, Bacon disparagingly notes that authors receive the status of dictators, "and not consuls to give advice."[9] The resulting damage to the natural sciences, he laments, "is infinite," and "the principal cause that hath kept them low, at a stay, without growth or advancement."[10] Between the willful ignorance of physical causes and the unquestioned authority of the ancient Greeks and the Christian church, the advancement of knowledge at the dawn of the seventeenth century was at a virtual standstill. Throughout his works, Bacon railed against unquestioned authority, which he argued is the most pernicious enemy of all.

Bacon set out to transform the state of knowledge with a new tool, or organon, for the advancement of knowledge. Real learning, according to Bacon, had been all but halted by unquestioned authority, be it religious or political. The result of this was that all progression of knowledge had ceased. Bacon's plan for the restructuring of human knowledge would not simply challenge what had been unquestionable, it would change the entire focus of the enterprise. His radical idea was that instead of looking up to heaven for mystical answers to divine phenomena, humankind must look to the earth and seek out the physical causes of natural phenomena. Bacon argued that there was only one way of seeking out and discovering truth: by deriving "axioms from the senses and particulars, rising by a gradual and unbroken ascent, so that it arrives at the most general axioms last of all. This is the true way, but as yet untried."[11] This formed the basis of his inductive method, and it would remain the principal method of scientific inquiry for centuries after his death. Bacon's views on the proper role of religion in the state are very much linked with his exhortations to advance knowledge of the natural world. The dethroning of religion from its place of dominance in matters of philosophy echoed through the following centuries and became perhaps the most controversial hallmark of modern science, which, embarrassingly enough, persists to this day.

Heated debate over the proper relationship between religion, science, and politics is hardly new to the twenty-first century. Natural and political philosophers alike have wrestled with this question since Bacon first proposed the separation of religious concerns, not only from scientific concerns but from political concerns as well. He reasoned that while religious extremism leads to violence, and politics all too often leads to the promotion of self-interest and the interests of the few over the good of the many, a focus on scientific accomplishments could ameliorate both and lead to political peace. "Of Unity in Religion," a vignette appearing in Bacon's *Essays*, helps illustrate Bacon's stance on the role religion should play in politics. Bacon considered the religious controversies of the day to be the result of extreme religious views, and in this essay he wrote that where the bounds of religious unity are concerned, the goal of extremists is war. For "certain zealants," he writes, "all speech of pacification is odious. [. . .] Peace is not the matter, but following and party."[12]

John Locke would make a similar argument in the *Letter Concerning Toleration*. Locke understood Bacon's main point in his literary dialogue *An Advertisement Touching a Holy War*. Mixing religion with politics leads to violence, hence wars for the sake of religion should neither be encouraged nor condoned.[13] The *Advertisement* is an important contribution to the debate on the role of religion vis-à-vis the state, its politics, and its foreign policy. Ultimately, the dialogue is concerned with the political consequences of those all-too-frequent companions, religious fundamentalism and military zeal. Bacon's ideal of a moderate Christianity is reflected in many of his works. Most notably among these are *The New Organon, The Advancement of Learning, Essays*, and *The New Atlantis*, a utopian fable about an advanced, peaceful island community where natural science holds sway. Throughout these works run the theme of transforming human knowledge and philosophy, moving both away from a dependence on received doctrine, and toward a separation of religion and political majority from scientific inquiry.

Bacon thought that we can maintain religious faith but not let it interfere with scientific investigation. Religious faith would continue to inform the purview and parameters of scientific activity throughout the Enlightenment era, as it does to some degree to the present day. Yet Bacon knew that it was proper to give to faith that which is divine. Give to science that which belongs to the natural realm.

The misevaluation of human capacity is a significant component of Bacon's argument, and his connection of power to knowledge is the single most misunderstood and misquoted element of Bacon's philosophy. Knowledge and power are not one and the same for Bacon, but they are closely related. He stipulated that "nature to be commanded must be obeyed." In short, in order to obey nature, that is, to discover the laws by which nature operates and thus produce works that will benefit mankind, "human knowledge and human power meet in one; for where the cause is not known the effect cannot be produced."[14] Knowledge and power are thus linked in the sense that human power is needed to increase the store of human knowledge, and *not* in the sense that scientific knowledge leads directly to some kind of power to dominate mankind, ideologically or politically, as much as many old twentieth-century postmodernist writers would have wished for you to believe. As Bacon emphasized,

> Man is but the servant and interpreter of nature: what he does and what he knows is only what he has observed of nature's order in fact or in thought; beyond this he knows nothing and can do nothing. For the chain of causes cannot by any force be loosed or broken, nor can nature be commanded except by being obeyed.[15]

Indeed, Bacon makes clear that his purpose is not to replace one authoritative dogma for another, but rather to discover the laws of nature as they really are. What is needed to advance scientific knowledge is not for the scientist to be a glory seeker. Instead, "the true and legitimate humiliation of the human spirit" and a love of truth are the necessary character traits true philosophers of nature should hold.[16] His plan for natural science relies neither on individual bursts of genius nor random chance. A set and sure method is the only thing that will allow true scientific progress to be made. Bacon emphasized the importance of this method being both public and collaborative, so that those following it will be able to detect errors committed by others. These errors can then be corrected, and the stock of knowledge increased.

This "interpretation of nature" was to be carried out by induction. He knew that this method, as he put it, is "extracted not merely out of the depths of the mind, but out of the very bowels of nature."[17] In other words, the method proceeds from facts unearthed to the formation of laws of nature. Bacon's induction is not "simple enumeration," which he calls "childish." Rather, it is a method for the "discovery and demonstration of sciences and arts," which "must analyze nature by proper rejections and exclusions; and then, after a sufficient number of negatives, come to a conclusion on the affirmative instances—which has not yet been done or even attempted."[18] Hence this method would lead to a gradual building of knowledge and, he hoped, prohibit those who would engage in scientific study from jumping immediately from particular instances to false assumptions.

The method is not foolproof, and Bacon was aware of it. One of its most significant drawbacks is that the senses are prone to deceive us. The senses can lie, they can give us no information where information exists, or they can give us false information altogether. And since our senses are generated internally, we are in a way preprogrammed to see either what we wish to see, or see what we expect to see. Our preconceived notions can harm a proper interpretation of nature and are the bases of what Bacon called the "idols of the mind": the "false notions which are now in possession of the human understanding, and have taken deep root therein."[19] As deeply rooted prejudices, they are not possible to overcome completely. We can, however, be "forewarned of the danger" and "fortify [ourselves] as far as may be against their assaults."[20]

Seventeenth-century natural scientists and political philosophers agreed. Chemist Robert Boyle argued that a proper scientific conception of nature should be precisely defined, so that all may agree on a common foundation on which progress can be built. Political philosopher Thomas Hobbes would later note that imprecisely defined concepts only harm the advancement of scientific knowledge. More than anyone else, Bacon inspired his scientific and philosophical descendants to develop a language of science amenable to the discovery of natural laws. The sciences began to diversify, taking on their own languages and terminology, including the study of physics. As the inventor of physics Isaac Newton demonstrated in *The Principia* later that same century, that language is mathematics.

When dealing with politics, however, "it is not the lie, that passeth through the mind, but the lie that sinketh in, and settleth in it, that doth the hurt," as Bacon so eloquently stated it.[21] His following words are worth quoting in full. Why? Because they are not only true in terms of religious interference in scientific matters, but they are also true when it comes to the relationship between natural science and politics. And since in the United States various adherents of one religion or another all too often consider themselves to be self-appointed political mouthpieces when it comes to any scientific advancement they don't "like," I offer another serving of Bacon:

What a man had rather were true he more readily believes. Therefore he rejects difficult things from impatience of research; sober things, because they narrow hope; the deeper things of nature, from superstition; the light of experience, from arrogance and pride; . . . things not commonly believed, out of deference to the opinion of the vulgar. Numberless, in short, are the ways, and sometimes imperceptible, in which the affections color and affect the understanding.[22]

Individual judgment rather than the slavish following of received opinion is the key to Bacon's philosophy. According to him, genuine consent does not flow from bowing to doctrine. Merely following the words of previous masters is a sign of inadequacy in the realm of knowledge, not confirmation of rectitude. Rather, "true consent is that which consists in the coincidence of free judgments, after due examination."[23] This trope of free consent is a crucial one. It would be adapted from the realm of scientific inquiry and enter political thought to become part and parcel of the basis of classical liberalism.

After Bacon's death, scientists applied his method to the study of the natural world, and philosophers adapted tenets of it in their explications of the human condition and political thought. By the end of the century, John Locke would apply many core Baconian tenets to the realm of political philosophy, and in so doing he influenced the form modern liberal political theory would take.

The Second Treatise of Government is John Locke's most politically influential work. His formulation of a government based on consent and the legitimacy of questioning authority, which includes the authority of any majority, reflects Bacon's own stance on natural science in *The New Organon*, in which Bacon cautions us that "true consent is that which consists in the coincidence of free judgments, after due examination."[24]

In the first chapter of the treatise, Locke states that *political power* is "*a Right* of making laws . . . and this only for the Publick Good."[25] Locke's intellectual construct of the state of nature—the pre-political condition humans inhabited before deciding to live and work together for mutual benefit—is an important starting point. For "the *State of Nature* has a Law of Nature to govern it, which obliges every one: And Reason, which is that Law, teaches all Mankind, who will but consult it, that being all equal and independent, no one ought to harm another in his Life, Health, Liberty, or Possessions."[26] The language he uses is vital: "a Law of Nature to govern it." With this, Locke lays the foundation for connecting the laws of nature to a government based on consent, which itself must obey those laws. In *The Second Treatise of Government*, reason is the law of nature. And since all are capable of reason, or as he puts it, "we are born free as we are born rational,"[27] Locke is able to posit both equality and natural freedom. "Within the bounds of the Law of Nature," all may act as they see fit.[28]

This was an extraordinary and crucial move. And it was certainly not a theory of democracy. It was something entirely new. Locke's discussion of political power in *The Second Treatise* is deeply connected to his ideal of individual liberty. He did nothing less than adapt the language of natural science in general, and Newtonian physics in particular, to formulate his ideas of power, liberty, and necessity. *The Second Treatise* links these concepts to government. As he explained, "The *Natural Liberty* of Man is to be free from any Superior Power on Earth, and not to be under the Will or Legislative Authority of Man, but to have only the Law of Nature for his Rule."[29] The "Law of Nature" here is being used in the style of Isaac Newton's three laws of motion. In *The Second Treatise*, our natural state (our "natural liberty") is to be free from the will of another. It is to be free from "Legislative Authority." These ideas are wholly influenced by natural science. And as we shall see, these newly formed political ideas share an ethic that is parallel to that of natural science.

For Locke, our natural liberty in the state of nature is not lost when we form civil society: "The *Liberty of Man, in Society*, is to be under no other Legislative Power, but that established, by consent, in the Commonwealth, nor under the Dominion of any Will, or Restraint of any Law, but what the Legislative shall enact, according to the Trust put in it."[30] Locke's reliance on consent reflects Bacon's stance in *The New Organon*, where the latter argued that true consent—not arbitrary authority—should be the only real basis of scientific agreement.

Freedom in the state of nature is the power to act only within the law of nature, with reason as a constraint. In any society this freedom is curbed according to an agreed upon rule of law. Locke argues that the law is not meant to take away freedom, but enhance it to the benefit of society. Thus he is able to state that "*the end of Law* is not to abolish or restrain, but *to preserve and enlarge Freedom*: For in all the states of created beings capable of Laws, *where there is no Law, there is no Freedom.*"[31] Again, the language he chose is of paramount importance. Liberty is defined as the power "to be free from restraint and violence from others which cannot be, where there is no Law: But Freedom is not, as we are told, *A Liberty for every Man to do what he lists*: (For who could be free, when every other Man's Humour might domineer over him?)"[32]

Thus liberty stems from an individual's power to do or to forbear. Only now has this individual power been writ large. First to the level of society, and now to the formation of government. Yet at the same time the right of the individual is never lost. Locke makes it crystal clear that we are each born with rights. They are inalienable. A democracy cannot, nor has it ever successfully tried to make such a claim. Democracy is "rule by the people." A majority of 50 percent of the population plus one person constitute the ruling majority. The concepts of individual liberty and inalienable rights do not, nor could they ever, apply in such a political schema.

Locke argues that reason enables individuals to conclude that forming a government based on consent will preserve their lives. The preservation of their individual rights, themselves stemming from the laws of nature and natural liberty, plays a key role in Locke's political theory. In *The Second Treatise*, citizens elect representatives via democratic choice to rule their "Commonwealth," a form of government characterized by a separation of powers (Executive, Legislative, and "Federative") and described by Locke as "any independent community which the Latines signified by the word Civitas," and in which the elected Legislative branch has the power to make laws by which it is itself bound.[33] As Locke puts it:

> The great end of Men entering into Society, being the enjoyment of their Properties in Peace and Safety, and the great instrument and means of that being the Laws establish'd in that Society; the first and

fundamental positive Law of the all Commonwealths, is the establishing of the Legislative Power; as the first and fundamental natural
Law, which is to govern even the Legislative itself, is the preservation of
the Society and (as far as will consist with the publick good) of every
person in it.[34]

Thus far, Locke, like Thomas Hobbes did before him, has transcended
custom and traditional authority by denying the supremacy of the ecclesiastical over the political, by positing that humans are essentially equal,
and by arguing that we are capable of reason. But while Hobbes's solution to the maintenance of his subjects' liberty was the appointment of
an absolute sovereign, Locke's remedy is his "Commonwealth," a government characterized by representation, the separation of powers, and the
rule of law. Both Hobbes and Locke relied on the categories, language,
and ethic of modern natural science as first envisioned by Francis Bacon
to demonstrate that humans are capable of reason, and that they possess
the natural capacity of political self-determination *via mutual consent.*

As we can see clearly, Locke's individuals enter into civil society and
form the commonwealth to protect their right to life, liberty, and property. The government is based on consent—that indispensable element
of Bacon's natural science—and the only basis of "any *lawful Government* in the World," according to Locke.[35] As such, "against the laws there
can be no authority." Locke establishes the existence of laws of nature in
such a way that no authority could legitimately contradict them, be that
authority ecclesiastical, temporal, or military. Were this not the case, any
powerful authority could violate the original purpose of the commonwealth, which includes the safety, security, and the protection of the people's inalienable rights. This brings us to the move Thomas Hobbes would
never have dared to consider. If the "supreme executive power" of Locke's
commonwealth were to "neglect and abandon" the charge of preserving
the rights of the people, rights that are naturally ours, the government
may be dissolved by the people themselves.[36]

Once dissolved, "the people are at liberty to provide for themselves,
by erecting a new Legislative, differing from the other."[37] The process by

which this is determined is as powerful as it is simple: "*The people shall be Judge.*"[38] Of course, Locke is confident that a government such as he has constructed, one that is in the name of the people for their preservation of their lives, liberties, and estates, should have very little need for rebellion. After all, the government is designed to be self-corrective from the very beginning. As he states, "every little mismanagement in publick affairs" is not enough to foment revolution; only "a long train of Abuses, Prevarications, and Artifices, all tending the same way, make the design visible to the People."[39]

Locke's innovation is not just the right of humans with equal capacity for reason to consent to establish a government for themselves; it is the perpetual right to question such authority in perpetuity. Just like the ethic and practice of modern natural science still demands. This political innovation was radical, and it formed the basis of the classical liberal tradition, a tradition that we live by and still hold dear. As Locke explains it,

> Tyranny is the exercise of power beyond Right, which no Body can have a right to. And this is making use of the Power any one has in his hands; not for the good of those, who are under it, but for his own private separate Advantage. When the Governour, however intituled [*sic*], makes not the Law, but his Will, the Rule; and his Commands and Actions are not directed to the preservation of the Properties of his People, but the satisfaction of his own Ambition, Revenge, Covetousness, or any other irregular Passion.[40]

Succinctly put, wherever law ends, tyranny begins.

Grounded in reason and drawing on the language and ethic of natural science, Locke, following Francis Bacon, went beyond the latter's exhortation to question religious and intellectual authority, and declared that humans have the right to resist established political power. The progression of his argument is elegant. Since all are capable of reason, which is the primary "law of nature," we possess certain inalienable rights, based on the laws of nature that no one can violate. Included in these rights is the power to establish government, to limit its sway, and to sanction the formation of laws to which the government must itself be subject.

Since these rights are based in the laws of nature, which no authority can eradicate or refute, the people have a right to question any authority that breaks these laws. Hence the ethic of natural science became the foundation onto which the political procedures of republics were built. No democracy has ever had such an ethic.

It is very important to note that Locke was more concerned with establishing a stable, institutional means of preserving the natural liberty of individuals than he was in creating a "perfect" society. The difference between the two approaches is significant. While the latter approach involves an attempt to identify and remove the causes of dissent and injustice, the former endeavors to establish a form of government designed to be self-correcting and flexible in the face of dissent and injustice. This difference is what separated the Continental philosophical tradition stemming from René Descartes from the philosophical tradition inspired by Bacon and the scientific revolution. In other words, it formed the basis of classical liberalism, which has at its center the core ideas of individual freedom, inalienable rights, and the legitimacy of political resistance.

Locke wasn't interested in theorizing utopia. Instead, he was concerned with incorporating a new way of thinking about the natural world into his answer to that principal question of political philosophy: *How ought we to live?* For Francis Bacon at the outset of the seventeenth century, as in Locke's "Commonwealth," and as it is in the activity of natural science today, flexibility in the face of new information is crucial.

The activity of science cannot be reduced to mere rote method, much less an ideology. It embodies an ethic of unceasing questioning not only of scientific authority, but also of intellectual, political, or religious authority. The activity of science is ongoing and continuous, not driven by a desire to find comprehensive Meaning, Absolute Truth, or The Answer to the vagaries of life. Searches such as these fall under the purview of religion and metaphysics. Rather, the activity of science involves the endeavor to seek answers to specific questions posed about the physical universe. The goals of science, as Bacon first explained, are to pursue knowledge of the natural world, and to use this knowledge to benefit and aid humankind.

The legacy of Bacon and Locke, like the legacy of natural science, did not end with the seventeenth century. The advancement of natural science went hand in hand with the expansion of the idea that people have a right to govern themselves, and that political resistance is a legitimate act. Both formed the basis of classical liberalism. The Lockean and scientific elements within the American Declaration of Independence constitute one extraordinary example of how these principles continued to shape the modern world after the seventeenth century. The beginning of the Declaration deserves to be reread. Why? Because it deserves to be read *this* time with an eye to the influence, the language, and the ethic that the inception of natural science gave it. The italics for emphasis are mine:

> When in the Course of human Events, it becomes necessary for one People to dissolve the Political Bands which have connected them with another, *and to assume among the Powers of the Earth, the separate and equal Station to which the Laws of Nature and of Nature's God entitle them,* a decent Respect to the Opinions of Mankind requires that they should declare the causes which impel them to the Separation.
>
> We hold these Truths to be self-evident, that all Men are created equal, that they are endowed by their Creator with certain unalienable Rights, that among these are Life, Liberty, and the Pursuit of Happiness—That to secure these Rights, Governments are instituted among Men, deriving their just Powers from the Consent of the Governed, that whenever any form of Government becomes destructive of these Ends, it is the Right of the People to alter or to abolish it, and to institute new Government, *laying its Foundation on such Principles,* and organizing its Powers in such form, as to them shall seem most likely to affect their Safety and Happiness. Prudence, indeed, will dictate that Governments long established should not be changed for light and transient Causes; and accordingly all Experience hath shewn, that Mankind are more disposed to suffer, while Evils are sufferable, than to right themselves by abolishing the forms to which they are accustomed. But when a long Train of Abuses and Usurpations, pursuing invariably the same Object, evinces a Design to reduce them under absolute Despotism, it is their Right, it is their Duty, to throw off such Government, and to provide new Guards for their future Security. Such has been the patient Sufferance of these Colonies; and such is now the Necessity which con-

strains them to alter their former Systems of Government. The History of the present King of Great-Britain is a History of repeated Injuries and Usurpations, all having in direct Object the Establishment of an absolute Tyranny over these States. *To prove this, let Facts be submitted to a candid World.*

These powerful words forged a unique nation with a unique objective. And these words were themselves forged in the ethic of modern natural science.

In the twentieth century, critics of natural science from both extreme ends of the political and ideological spectrum tragically overlooked the emancipatory and radical elements of natural science and its influence on political thought. Even some of the most influential of these critics, such as Theodor Adorno, Max Horkheimer, Michel Foucault, and their philosophical descendants, the self-styled postmodernists, based their ideology not on natural science as it actually evolved since Francis Bacon, but rather on metaphysical assumptions and so-called *a priori* principles. The influence of such a misguided endeavor was not only to eclipse the vast importance of Bacon's work, but also to draw a false image of the activity of science simply to benefit its critics. The result of their misconceptions was that the very phenomenon they hoped to counter they in fact resurrected from centuries ago. Their numerous critiques of various so-called "hegemonies" of one mode of thought or another took on new forms in their mouths. Unlike Bacon's interpretation of nature, which sought to mitigate the dangers of the idols of the mind, these critics of natural science were satisfied to first proclaim and then transfer its supposed dogmatic authority to themselves. Claims that hovered around the ideas that natural science succeeded only in of the wake of European imperialism were strong. Whether the wake was military or cultural or economic or technological, it very quickly became impossible to refute postmodern doublespeak without leaving oneself open to accusations of promoting "imperialism" oneself, and hence summarily dismissed. Reason, evidence, and method all became epithets in such circles.

The scientific method is self-corrective. It demands that we use our senses and our capacity for reason. And it demands that we take the world on its own terms and not conform our perceptions to accord with our preconceived notions, no matter how cherished they may be. These elements form a powerful base on which sits seventeenth-century Enlightenment thought, the foundation of the modern era. This is the legacy of Francis Bacon.

Attacks on natural science that misunderstand this will lead not to an emancipation from political, religious or cultural dogmatism, but rather to a condition of apathy toward political resistance itself. Without the ethic of modern natural science, we will be left only with what Bacon argued against: competing belief systems, such as religious factionalism or political extremism that have historically led to violence; appeals to authority of one sort or another; or slavish adherence to the discourse of the moment. Let us use our own judgment in the realm of scientific inquiry as in the realm of civil action. Let us be our own masters. Let us not depend on various authorities to think for us. Let us think for ourselves.

CHAPTER 6

THE LEFT, SCIENCE STUDIES, AND GLOBAL WARMING

MARGARET C. JACOB

N one of the seventeenth-century founders of modern science—Bacon, Descartes, Boyle, Hooke, and Newton—could have imagined the peculiar world of contemporary anti-science polemics. They thought that having some control over nature, particularly over human health—however desired—was an almost unimaginable goal. They believed that certain methods—experience, experiment, inductive and deductive thinking, and mathematical reasoning—might provide access to the intelligibility of nature. Only when their ideas got in the way of religious doctrines taught by the clergy—or by their own religious scruples—did natural philosophy come under attack.

At the University of Utrecht, during the 1630s and well beyond, the followers of the philosopher/theologian Gisbert Voetius would have run Descartes out of town, more precisely out of the Dutch Republic, if only the Voetians had the backing of magistrates willing to persecute. Voetius rightly concluded that Descartes's mechanical philosophy could not be reconciled with Aristotle. Shrewdly, fellow clerics offered other, skillfully damning arguments. They argued from the pulpit, for example, that if the educated elite took up a mechanical understanding of man and nature they would renege on their obligations to the poor. Always the objections to Cartesianism included the fear that it would lead to materialism, and hence atheism.

Newton thought that Descartes's system would do just that, and devout English (if heretical) Protestant that he was, Newton was intemperate on the subject of atheism.[1] In seventeenth-century Holland, as in

England, religious passions and interests infiltrated the discussion when it came to assessing the worth of science, or what the age called natural philosophy. The protagonists thought little about the implications of their philosophical positions for the future or for the nature of state power.

Today, guidance for such a long-sighted analysis should be able to draw from the writings of the discipline "science studies," and from Bruno Latour, one of its first practitioners. In *We Have Never Been Modern*, Latour analyzed the motives of two seventeenth-century natural scientists, Thomas Hobbes and Robert Boyle, and assessed the meaning of their respective positions about science.[2] Hobbes opted for mathematics as its anchor, while Boyle articulated—with singular clarity—the experimental method. The positions being staked out, according to Latour, were a "*mere* distinction between two regimes of representation" two different ways of doing science, both of equal merit.[3] Both were strategies for securing power, neither had implications for the kind of polity that would arise out of the turmoil of the midcentury revolution, or so he thinks.

Latour either does not know or does not care that had Hobbes's understanding of science triumphed royal absolutism would have rested far more secure. Boyle's account validated an independent sphere of witnessing unaffiliated with the state, in effect a version of civil society. His vision of scientific practice did little to enhance the power of monarchy. It valorized a space where *virtuosi* could experiment with air pumps or collect samples from nature without regard to the royal will. By contrast, Hobbes opted for mathematical learning because it provoked few controversies and thus enhanced political stability. Had Hobbes won against Boyle, independent intellectual life, without state sponsorship, would have contracted. The implications for the development of the institutions of liberal democracy in England are obvious.

If the relativism in the Latourian method failed to reveal the historical stakes when applied to the English Revolution, perhaps it would work better when put to use in the here and now. Into the contemporary crisis provoked by the deniers of global warming, we might reasonably expect that academic commentators who study the nature of science (like Latour), as distinct from those doing natural science in the field or laboratory, would enter the discussion and sort through the polemical debris. There are approximately sixty programs or departments in the United States where science is studied for its history or its philosophical meaning. The British academy also excels in the history and philosophy of science, and smaller outposts can be found in the Netherlands and France. In a few of them, generally known under the Anglo-American label "science studies," orthodoxy requires that no explanation for the historical development of a science can invoke its relative truthfulness as the reason why contemporaries chose a particular approach or embraced a body of factual material, since this would be considered positivism. It follows: the truth or falsity of global and anti–global warming science is irrelevant.

From science studies, therefore, guidance through the minefield of global warming and its deniers is in short supply. Let me return again to the ever-fashionable Latour, once found at Sciences Po in Paris, and now emeritus of that institution. His critics accuse him of epistemological relativism, a charge that amounts to the ultimate heresy against the academic pursuit of truth. The accusation must have hurt enough for him to answer it in an essay that appeared in *Critical Inquiry* in 2004, during the Bush administration when the anti–global warming, anti–stem cell crowd were riding high. Latour, to his credit, wondered: Had the science studies movement, with its recourse to social construction and discourse analysis, made this crowd's life easier?[4] He could not have foreseen what was coming in 2016–2017.

Urging his compatriots in science studies in 2004 to take stock, Latour asked if the questioning of scientific facts had led to our forgetting "a realism dealing with what I will call matters of concern, not matters of fact. The mistake we made, the mistake I made, was to believe that there was no efficient way to criticize matters of fact except by moving away from them and directing one's attention toward the conditions that made them possible."[5] Facts, Latour is saying, are not the real issue: "Matters of fact are only very partial and, I would argue, very polemical, very political renderings of matters of concern and only a subset of what could also be called states of affairs." Thus liberated from the burden imposed by an excessive zeal for facts, in 2004 Latour was free to question any narrative about the way the natural world might be, whether it be mathematical, empirical, or, of course, historical.

Then came Donald Trump. Suddenly a tragic reality crashed into the relativism so carefully cultivated by Latour and his followers. His response? He found little to separate Trump from the Clintonian alternative: "The real tragedy, though, is that the others [the Trumpites] live in a bubble, too: a world of the past completely undisturbed by climate change, a world that no fact, study, or science can shake."[6] No amount of the factual or the scientific can disturb the people who live in Trump's bubble. According to Latour, the electoral alterative was also really no better: "The opposition between Clinton and Trump illustrated this rather well: both occupied their own bubbles of unrealism. For now, the utopia of the past has won out. But there's little reason to think that the situation would be much better and more sustainable had the utopia of the future triumphed instead." Had Latour been allowed to vote in the November 2016 election, he probably would have stayed home. With the unfailing instinct of the contemporary European Left for irrelevance, Latour proclaims that there really was no difference between Clinton and Trump: as he puts it, "Thus, two utopias: a utopia of the future confronting a utopia of the past."[7]

Latour's smugness, evident in November 2016, within little more than a year has given way to advocacy, to precisely what Clinton wanted Americans to do—to radically and immediately address the reality of climate change and do something about it. And Latour now thinks that the scientists need him: "So work I had done many years ago—saying we have never been modern—is now being vindicated. Everyone agrees we will never modernize the planet."[8] The reality of climate change has put a new spring in his step and validates, he believes, his repudiation of the experimental direction taken in the time of Boyle.

Latour's major complaint always centered on the narrative of Western modernity. Contrary to what many historians and the general public might believe, Latour proclaimed modernity an illusion, and the source of the false dichotomy we imagine between "the social" and "the natural." In effect, he blames the experimental science that emerged out of the work of Boyle and Newton for vindicating the exploitation of nature and the resulting danger to our planet. The science that emerged out of the seventeenth century rendered nature passive, subject to mechanical laws, and awarded agency only to humans. The hubris of science, set in place in the seventeenth century, has allowed the arrogance that is wrecking the planet.

Back in the 1990s, Latour wanted to escape the dichotomy between the natural and the social, opting instead for the posture of a "non-modern." He wished to award equal agency to all entities—things, objects, humans, and beasts—and in the process find a definitive way out of the social constructionist/anti-realist vs. realist/positivist confrontation that so plagues the discipline of science studies and its critics. He wanted to rescue passive nature from being subjected to socially constructed maxims posing as facts or laws. One adoring acolyte found Latour's philosophical move so brilliant as to make him "the Galileo of metaphysics, ridiculing the split between the supra-lunar world of hard scientific fact and the sub-lunar world of human power games."[9]

Given all the hype, how might we expect a Latourian practitioner to address the naysayers about climate change? Or—to pose the question in an historical context—how would he explain the seventeenth-century Dutch Calvinist minister who thought Cartesianism would lead to the neglect of the poor? Or Newton's fear that Descartes's science would lead inexorably to atheism? The Latourian might have recourse to human power games as the long and short of what motivated seventeenth-century critics of science or natural philosophy. The sincerity of their religious beliefs would be beside the point; the status of one's natural philosophy would matter, as would the maintenance of clerical power. Very little outside their immediate power needs would factor into an analysis of the positions adopted. Larger motives with future geopolitical implications would be beside the point.

If it had the methodological tools, science studies could do better than this weak analysis. If it could assess the denial of global warming within the context of political power and do the research to document the fiscal interconnectedness of corporations with the deniers, this relatively new discipline would secure its future. First the empirical research would take a close look at who finances the anti–global warming initiatives. Following the money would lead to evidence of naked self-interest posing as serious intellectual inquiry. A power game, indeed. One with geopolitical implications that pose a mortal threat to democratic societies.

Starting from the readily available knowledge, based upon science, and found, for example, at science historian Spencer Weart's website, *The Discovery of Global Warming*,[10] science studies could examine the political implications to be drawn out of the denial of global warming. Do we think for a moment that democratic politics will flourish in a situation of extreme water shortage or in the face of life-threatening pollution? The financial interests that work to deny global warming would hold monopoly rights over land and sea as well as energy resources.

Historically informed, contextual approaches might enable us to disaggregate the motives of the present-day crowd of anti-science polemicists. Each must be seen in relation to their broader interests and commitments. These can range from the will to make a profit—the environment be damned—to believing that America first, or Russia first, is essential if natural resources are to be carefully monopolized.

Finally, readers without any direct experience of science studies or the writings of Latour, might well be asking, how did a field once so promising become laden with dead-end philosophical disputes and positions? Nothing about this situation can be understood without reference to the Cold War. To put it as my generation, being trained in the 1960s, said it: How would we write a history of Western science after World War III? The whole notion of the control of nature, the Baconian discourse about the improvement of man's estate, would become unthinkable, even immoral.

The enormous power of postwar American science and technology (in constant competition with the Soviets) demanded deconstruction, an exposure of the power relationships embedded within the entire structure of Western science, beginning at its seventeenth-century origins. A classic intellectual move, derived from the ancient Greeks, required a turn toward skepticism, more precisely relativism. Science Studies was the result—but so too was the license to question any scientific finding. No one thought until very recently that the anti-science forces would receive powerful political backing and make their way into actual policymaking, that they would, for one example, dismantle environmental regulations.

Only relativism justifies advocating the teaching of creationism and evolution in the same curriculum, or teaching that global warming is a hoax.[11] Only a commitment to furthering democratic values and the science that would ensure them can take up the challenge posed to our future by the climate-warming deniers. Only such a commitment would allow an interrogation of their interests and agenda and propel research into their economic and political entanglements. New non-Latourian voices are urgently needed.

CHAPTER 7

BETRAYING THE FOUNDERS' LEGACY: DEMOCRACY AS A WEAPON AGAINST SCIENCE

BARBARA FORREST

Building a new country is hard work. Plato had interesting ideas about the model city-state, but his plans never got beyond the thinking stage. The American Founders, on the other hand, faced with the opportunity—indeed, the necessity—to do more than dream up theoretical blueprints for a new political system, actually constructed one. They built a democracy, and not just any kind. They wanted something that had never existed: a secular, constitutional democracy. Church and state would be separate, each having power over its respective realm and no more. This would be a fortunate arrangement. Religious persecution, a European tradition that American settlers brought with them, had plagued the colonies for more than a century. But the Founders were also decidedly pro-science. No longer would the ecclesiastical institutionalization of scientific error—Galileo's conviction was little more than a century old—obstruct the pursuit of knowledge that would make life more secure and enjoyable. The natural sciences would be the engine of America's technological and economic advancement. In the new United States of America, science and democracy would progress together.

The historical record since the Founders formalized their vision in the United States Constitution shows that the dual promotion of science and democracy has been imperfect. The American experiment has been marred by honest mistakes, sheer stupidity, and gross inhumanity. Nevertheless, our secular democracy has enjoyed the benefits of science to an extent that the Founders never conceived. So why, from the twentieth century until now, the twenty-first, has the Republican Party been engaged in a sustained attack on science?

The answer to this question lies partly—but significantly—in the Republican Party's absorption of the Religious Right into its political operations. Science journalist Chris Mooney, in his well-researched book *The Republican War on Science*, explains this development, which began during the administration of Ronald Reagan and shifted into high gear under George W. Bush: "Reagan brewed a political concoction—equal parts big business and religious conservatism—that proved highly toxic to the role of science in government."[1] Bush continued the attack on science, which, although rooted in the Republican Party's conservative ideology, cannot be explained by that alone: "During its rise to political triumph and domination of the Republican Party, the modern conservative movement has relied heavily on two key constituencies with an overriding interest in the outcomes of scientific research in certain areas: industry and the Religious Right. . . . Religious conservatives . . . seek to use science to bolster their moralistic agenda."[2] Under the administration of Donald Trump, the attack on science has continued with a vengeance, supported by the same anti-science constituencies.[3]

While the Republican Party's pro-business-and-industry constituency is also involved in anti-science efforts, this article focuses on the role of the Religious Right in the attack on science. Rather than emulating the Founders' support for science, Republican politicians, allied with the Religious Right, are using the mechanisms of democracy in order to undermine it.

THE FOUNDERS' SUPPORT FOR SCIENCE

I. Bernard Cohen's excellent book *Science and the Founding Fathers: Science in the Political Thought of Thomas Jefferson, Benjamin Franklin, John Adams and James Madison*, documents the Founders' integration of their enthusiasm for science with democratic politics. Many attended college and were schooled in the science of their day: "At Harvard, at Yale, at Princeton, at William and Mary, students were required to study mathematics and the principles of Newtonian science and were introduced to the new sciences being created or advanced in their own time."[4] Not only were future presidents John Adams, Thomas Jefferson, and James Madison well informed about contemporary science, but their scientific literacy also had a measurable influence on their intellectual and political lives.[5]

Even Founders who never attended college understood the benefits of science for democratic government. Foreseeing the need for scientifically educated citizens even before the American colonies declared their independence, Benjamin Franklin founded the American Philosophical Society in Philadelphia in 1743. He explained the importance of this organization:

> The first Drudgery of Settling new Colonies, which confines the Attention of People to mere Necessaries, is now pretty well over; and there are many [men] in every Province [colony] in Circumstances that . . . afford Leisure to cultivate the finer Arts and improve the common Stock of Knowledge. To such of these . . . many Observations occur, which . . . might produce Discoveries to the Advantage of some or all of the British Plantations [colonies], or to the Benefit of Mankind in general.[6]

The society was headquartered in Philadelphia, and Franklin stipulated that the membership there must include "a Physician, a Botanist, a Mathematician, a Chemist, a Mechanician [engineer], a Geographer, and a general Natural Philosopher [scientist]"; members in the other colonies were required to maintain a "constant Correspondence" and hold monthly meetings to discuss their scientific findings and technological achievements.[7] This shared information had to include, among other things, "New Methods of Curing or Preventing Diseases. . . . New Discoveries in Chemistry, . . . [and] New Methods of Improving the Breed of useful Animals"—in short, "all philosophical [scientific] Experiments that let Light into the Nature of Things, tend to increase the Power of Man over Matter, and multiply the Conveniencies or Pleasures of Life."[8]

Franklin, a self-educated scientist, became a Fellow of the Royal Society in 1756 and a foreign associate of France's Académie des Sciences in 1773.[9] Yet he never held the one office that wielded the political power to promote the institutional advancement of science. That task fell to his younger associates: George Washington, Adams, Jefferson, and Madison.

Like Franklin's, Washington's formal education ended in adolescence.[10] However, his 1783 *Circular Letter of Farewell to the Army*, announcing his retirement as commander-in-chief of the colonial army, reveals his awareness of the potential benefits of science and democracy to the country's future. America's foundation "was not laid in the gloomy age of Ignorance and Superstition," but rather during the Enlightenment, "when the rights of mankind were better understood and more clearly defined, than at any former period" and when "knowledge, acquired by the labours of Philosophers [who, in the eighteenth century, included natural scientists], Sages and Legislatures" could be "happily applied in the Establishment of our forms of Government."[11] Practicing what he preached, Washington put the latest agricultural science to use at Mount Vernon.[12] As his presidency ended in 1797, he urged Americans to establish "institutions for the general diffusion of knowledge" as "an object of primary importance."[13]

Cohen writes that the politically conservative Adams considered natural science "the highest form of human knowledge based on reason and experience."[14] A Harvard graduate who "received as good an education in science as was possible in America at that time," he regretted spending more time reading political philosophy than science.[15] Yet Adams understood that American cultural and economic development depended on science: "I must study Politicks and War that my sons may have liberty to study Mathematicks and Philosophy [science]. My sons ought to study Mathematicks and Philosophy, Geography, natural History, Naval Architecture, navigation, Commerce and Agriculture, in order to give their Children a right to study Painting, Poetry, Musick, Architecture, Statuary, Tapestry and Porcelaine."[16] To that end, in 1780 Adams cofounded the American Academy of Arts and Sciences.[17] Its charter emphasizes the importance of "Arts and Sciences" to "agriculture, manufactures, and commerce" and to "the wealth, peace, independence, and happiness of a people; as they essentially promote the honor and dignity of the government which patronizes them."[18] Adams clearly believed that democracy would benefit from his promotion of science.

Jefferson, a William and Mary alumnus and the most scientifically astute founding president, gave up science for politics only because of "the enormities of the times."[19] While vice-president of the United States, he was also president of the American Philosophical Society. His commitment to both science and democracy is reflected in the portraits of John Locke, Francis Bacon, and Isaac Newton that he commissioned for his State Department office; he was, as Cohen says, "surely the only president of the United States who ever read Newton's *Principia*."[20] Jefferson's command of the sciences proved politically significant when he commissioned the Lewis and Clark expedition, which he used to acquire scientific data and fossils.[21] In *Notes on the State of Virginia*, Jefferson combined a thorough presentation of scientific data with a vigorous defense of both democracy and religious liberty.[22] According to Cohen, his reference to the "laws of nature" in the Declaration of Independence was as closely tied to Newtonian science as to political theory.[23] Cohen argues that Jefferson's introduction to the Declaration was rooted not only in "philosophy and political theory, his reading about natural law and natural right, [and] his training and experience in civil law" but also in "his knowledge of science."[24]

Madison, a Princeton graduate, was at a disadvantage where science was concerned. Although Princeton students studied mathematics, basic physics, and astronomy, the school's early history was marked by constant complaints from administrators and faculty that they lacked funding for scientific equipment. Later in life, however, Madison became a science devotee through his friendship with Jefferson. Under Jefferson's tutelage, he became proficient enough to secure membership in the American Philosophical Society, along with Joseph Priestley, the famed British chemist who had discovered oxygen, thus indicating the advanced level of scientific literacy that Madison acquired. Cohen points out that metaphors drawn from the natural sciences even made their way into Madison's famous *Federalist* No. 10.[25]

As Cohen documents, "the Founding Fathers displayed a knowledge of scientific concepts and principles which establishes their credentials as citizens of the Age of Reason."[26] Their commitment to science was second only to their commitment to political and religious liberty. The low status to which science has fallen in the Republican Party, not to mention the astonishingly low level of scientific literacy, would shock them.

USING DEMOCRACY AGAINST SCIENCE

Comparing the Founders' enthusiasm for science with Republicans' current animosity toward it is a sobering exercise, especially since this animosity is in great part the result of the Republican Party's alliance with the Religious Right, about which Barry Goldwater complained in 1994: "Mark my word, if and when these preachers get control of the [Republican] party, and they're sure trying to do so, it's going to be a terrible damn problem. Frankly, these people frighten me. Politics and governing demand compromise. But these Christians believe they are acting in the name of God, so they can't and won't compromise."[27] According to Mooney, this unholy alliance began with Ronald Reagan's support for teaching creationism in the 1980s: "The Reagan administration's sympathies with creationism signaled a new development for the Republican Party and conservatism. . . . From this moment forward, many of the party's leaders willingly distorted or even denied the bedrock scientific theory of evolution . . . to satisfy a traditionalist religious constituency."[28]

Republican animosity against science has not been a constant in the party's history. A 2017 editorial in the journal *Nature* points out that "Republicans have historically been strong supporters of science. They led the effort in the 1990s to double the budget of the National Institutes of Health (NIH), and they enthusiastically support space exploration."[29] Moreover, not all Republican office seekers have shared this animosity. Former Utah governor Jon Huntsman, who at this writing serves as the United States ambassador to Russia, announced (actually, tweeted, in somewhat less than serious language) his acceptance of evolution and global warming during his 2011 campaign for the Republican presidential nomination: "To be clear. I believe in evolution and trust scientists on global warming. Call me crazy."[30] However, during the 2011 campaign, Huntsman was an exception whose support for science became a political liability. Most Republican candidates have dutifully lined up with the science deniers.

Republican politicians' anti-science stance is not always mere political cynicism; some are both theologically conservative and scientifically (perhaps willfully) misinformed. Former United States congressman Paul Broun (2007–2015), a Georgia Republican who, incredibly, is a physician and served on the House Committee on Science, Space, and Technology, labeled evolution and the Big Bang theory as "lies straight from the pit of hell" that were taught in order "to try to keep me and all the folks who were taught that from understanding that they need a savior."[31] Other such examples of low science literacy and dismissal of serious scientific issues by Republican lawmakers are easy to find.[32]

Whatever the motivation in individual cases, the amply documented fact is that the Republican Party and the Religious Right are using the mechanisms of democratic government against science. There is no better example than the state of Louisiana under the governorship of Bobby Jindal from 2008–2016, an example that should serve as a cautionary tale to the rest of the country.

LOUISIANA—A DE FACTO THEOCRACY

With Brown University degrees in both biology and public policy, Jindal was well-schooled in both science and government.[33] Yet the contrast with the Founders could not have been starker. As governor, he entangled religion and government to a degree that would have made them blanch. Positioning himself for national office as chair of the Republican Governors Association in 2012, Jindal was the embodiment of his party's anti-science animus.[34] To please his right-wing religious base, Jindal, who labeled himself an "evangelical Catholic" in order to identify with his evangelical Protestant supporters, signed laws designed to undermine science education and punish scientists in Louisiana.[35] Kenneth Miller—a biologist at Jindal's alma mater—correctly describes Jindal's misuse of his office: "In his rise to prominence in Louisiana, he made a bargain with the religious right and compromised science and science education for the children of his state."[36] Under Jindal, Louisiana became a de facto "general theocracy," a government in which "ultimate authority . . . vested in a divine law or revelation [is] mediated through a variety of structures or polities," the relevant structures and polities being the governor's office, the legislature, and school boards.[37] Understanding how this transformation was achieved, how its agents viewed their roles, and how they exploited democratic processes is essential as a warning to other states and the nation at large.

Jindal's Religious Right allies began promoting anti-science legislation immediately after his January 2008 inauguration. Among the first bills he signed was the Louisiana Science Education Act (LSEA), which disguises creationism as "critical thinking" and was promoted as a safeguard of "academic freedom."[38] Louisiana has long been an incubator for religiously inspired—and democratically enacted—anti-science legislation, having enacted the 1981 "Balanced Treatment for Creation-Science and Evolution-Science in Public School Instruction Act" requiring the teaching of creationism along with evolution in public schools. The US Supreme Court declared the act unconstitutional in 1987.[39] Written and promoted by the Discovery Institute (DI), an intelligent design (ID) creationist think tank, and the Louisiana Family Forum (LFF), a Focus on the Family affiliate, the 2008 LSEA permits public school science teachers to use pseudoscientific supplementary materials in instruction concerning "evolution, the origins of life, global warming, and human cloning."[40] Five years after he signed it, Jindal admitted the true intent of the law: "The Science Education Act . . . says . . . if the [local] school board's okay with that, [and] if the state school board's okay with that, [teachers] can supplement those materials. . . . I've got no problem if . . . a local school board says, 'We want to teach our kids about creationism, that some people have these beliefs as well, let's teach them about intelligent design.' ... What are we scared of?"[41] Moreover, in 2012, Jindal successfully pushed a school voucher law under which millions of dollars in public funding were siphoned off to almost two dozen Christian schools that teach young-earth creationism (a problem that is spreading to other states along with adoption of taxpayer-funded school vouchers).[42]

In partnering with the Discovery Institute to promote the LSEA, LFF also helped DI to advance its national anti-science agenda—which is just as religion-driven as LFF's—by facilitating its exploitation of the legislative process. A DI staffer described its promotion of anti-evolution bills directly to legislators, both in Louisiana and other states: "Discovery Institute ... is on the inside. ... We draft and amend academic freedom language, counsel lawmakers privately, testify publicly, and are otherwise intimately acquainted with the intentions behind and likely effects of academic freedom legislation."[43] LFF's procurement of ground-level access to legislators for an out-of-state creationist think tank, with Jindal as guarantor of their success, epitomizes the Republican/Religious Right strategy of using democracy against science.

An ultra-conservative Catholic, Jindal also signed the 2009 "human-animal hybrid ban" on behalf of the Louisiana Conference of Catholic Bishops (LCCB).[44] S.B. 115, which was enacted as Act 108, threatens scientists with ten years at hard labor for, among other things, creating embryonic stem cells "by introducing a human nucleus into a non-human egg," for example, replacing genetic material in a bovine ovum with a human nucleus.[45] (As noted above, the LSEA targets instruction in "human cloning," thus making it also an indirect attack on stem cell research.) Such techniques, which were being done in regenerative medicine research that was licensed in the United Kingdom in 2008, the year Jindal took office, had elicited strenuous objections from the Roman Catholic Church.[46] The LCCB celebrated the passage of S.B. 115 as "the first law of its kind within the United States to our knowledge and reaffirms Louisiana as a consistent pro-life state specifically pertaining in this case to embryonic stem cell research."[47] LFF supported the bill as well.[48]

Religious anti-science efforts in Louisiana reflect only a fraction of such activity nationwide. Collusion between Republican politicians and the Religious Right occurs in other states, to be sure; Texas is a prominent example.[49] However, one might wonder about the nature of the coalition between Protestant evangelicals and Catholics on science-related issues such as the "human-animal hybrid" research barred by Louisiana's S.B. 115. LFF's joining the LCCB in support of this law stems from its membership in a Protestant-Catholic coalition of "pro-life" (anti-abortion) groups in Louisiana.[50] The attack on stem cell research and related science reflects the nationwide, working partnership between evangelical Protestants and conservative Catholics that was formalized in 1994 in a document entitled "Evangelicals and Catholics Together" (ECT) and reaffirmed in the 2009 *Manhattan Declaration*.[51] As stated in ECT, this partnership was forged in recognition of their joint agreement (despite serious doctrinal differences) that "Christians individually and the church corporately . . . have a responsibility for the right ordering of civil society."[52] As part of the entire spectrum of culture war issues that the mission of this partnership comprises, ECT specifically targets that most democratic of American institutions in which most Americans receive their only formal instruction about both science and democracy: the public school. ECT signatories affirm that public education requires Judeo-Christian religion as its foundation: "In public education, we contend together for schools that transmit to coming generations our cultural heritage, which is inseparable from the formative influence of religion, especially Judaism and Christianity."[53]

The *Manhattan Declaration* (*MD*), an extension of the partnership forged in ECT, focuses specifically on abortion, same-sex marriage, and religious liberty. From the first of these issues, abortion, stems the *MD*'s opposition to scientific research that is deemed a threat to the "sanctity of life": "As predicted by many prescient persons, the cheapening of life that began with abortion has now metastasized. For example, human embryo-destructive research and its public funding are promoted in the name of science and in the cause of developing treatments and cures for diseases and injuries."[54] Invoking the example of Martin Luther King, the *MD* asserts the need for civil disobedience, vowing that "we will not comply with any edict that purports to compel our institutions to participate in . . . embryo-destructive research."[55]

The subtext of both ECT and the *MD* is their putative respect for democracy, but that subtext clearly assumes that orthodox Christianity is the necessary foundation of democracy. Thus, both documents reflect an understanding of democracy that is advantageously—or, depending on one's point of view, dangerously—one-sided. Although ECT "strongly affirm[s] the separation of church and state," it "just as strongly protest[s] the distortion of that principle to mean the separation of religion from public life."[56] Asserting incorrectly that religion "was privileged and foundational in our legal order," ECT is critical of the courts' purported "obsession" with the establishment clause of the First Amendment to the detriment of the free exercise clause.[57] It argues fallaciously that religion must thus be restored to its rightful place as the basis of American democracy: "The argument, increasingly voiced in sectors of our political culture, that religion should be excluded from the public square must be recognized as an assault upon the most elementary principles of democratic governance. That argument needs to be exposed and countered by leaders, religious and other, who care about the integrity of our constitutional order."[58]

The charge that religion is increasingly excluded from "the public square" is a straw man. Proponents of church-state separation outside the ECT-*MD* alliance do not advocate the separation of religion from public *life*; in fact, precisely because of the separation of church and state, religion in many forms is ubiquitous in the lives of almost all Americans. The objection of proponents of church-state separation is to the enshrinement of religion as law and public *policy*.[59] Nonetheless, despite its stated intention "to elevate the level of political and moral discourse in a manner that excludes no one and invites the participation of all people of good will," ECT ultimately declares that the United States is "a nation under the judgment, mercy, and providential care of the Lord of the nations to whom alone we render unqualified allegiance."[60]

The preamble of the *MD* likewise assumes Christianity as the bedrock of western civilization, giving it exclusive credit for virtually all of the cultural and political progress in western Europe and America: the abolition of European and American slavery; the establishment in Europe of "the rule of law and balance of governmental powers, which made modern democracy possible"; the women's suffrage movement in America; and the American civil rights movement of the 1950s and 60s, which was "led by Christians claiming the Scriptures and asserting the glory of the image of God in every human being regardless of race, religion, age or class."[61] The document is silent about the fact that the injustices that Christians indeed helped to eradicate had been institutionalized by other Christians. It is likewise silent about the fact that the abolition of American slavery, women's suffrage, and civil rights for racial minorities could not have been accomplished without the United States Constitution—a completely secular document—and the secular federal courts for which the Constitution provides.

Just as the *MD* attributes the progressive aspects of western civilization to Christianity, so it grounds religious liberty in no less a figure than Jesus himself—ignoring the fact that prior to the adoption of the First Amendment of the Constitution, the Protestant majority denied religious liberty to many of their fellow Americans, including (and arguably, especially) Catholics: "The nature of religious liberty is grounded in the character of God Himself, the God who is most fully known in the life and work of Jesus Christ. . . . Thus the right to religious freedom has its foundation in the example of Christ Himself."[62] Moreover, the *MD* asserts that the "decline in respect for religious values in the media, the academy and political leadership . . . also threatens the common welfare and culture of freedom on which our system of republican government is founded," thus reiterating the view that the civic freedoms of democracy require the underpinning of religious values, specifically, the religious values of its signatories.[63]

Both ECT and the *MD* are clearly predicated on the assumption that democracy must serve the goals of this coalition of ultraconservative Protestant and Catholic signatories. Veteran journalist Frederick Clarkson, who has researched and written about the Religious Right for decades, minces no words concerning this prospect as laid out in the *MD*: "The *Declaration* . . . proposes a form of theocratic Dominionism—the antithesis of religious freedom."[64] Clarkson has defined dominionism in a general sense: "Dominionism is the theocratic idea that regardless of theological camp, means, or timetable, God has called conservative Christians to exercise dominion over society by taking control of political and cultural institutions. The term describes a broad tendency across a wide swath of American Christianity. People who embrace this idea are referred to as dominionists."[65] Elsewhere, he highlights its more specific aspects: "Dominionists celebrate Christian nationalism, in that they believe that the United States once was, and should once again be, a Christian nation. In this way, they deny the Enlightenment roots of American democracy. ... They believe that ... the US Constitution should be seen as a vehicle for implementing Biblical principles."[66]

Clarkson also spells out the implications of the merger of the Religious Right and the Republican Party: "Dominionism is therefore a broad political tendency ... organized through religiously based social movements that seeks power primarily through the electoral system. Dominionists work in coalitions with other religious and secular groups that primarily are active inside the Republican Party. They seek to build the kingdom of God in the here and now."[67] The successful use of democratic processes to achieve the dominionist goals enunciated in ECT and the *MD* would result in a distinctly theocratic state in which both democracy and science are undermined.

DOMINIONISM LOUISIANA-STYLE

In Louisiana, Bobby Jindal's use of the power of his office was utterly consistent with dominionist goals. During his governorship, dominionists rose to an unprecedented level of influence (which, although no longer quite as strong as when Jindal was governor, has diminished little with his departure). Jindal's primary partner in the fusion of religion and government was his close political ally Rev. Gene Mills, LFF executive director and an ordained Pentecostal Assemblies of God minister who epitomizes "these preachers" who so distressed Goldwater.[68] Mills has worked for many years as both the executive director of LFF and legislative lobbyist for its religious agenda, a role in which he has continued since the 2016 election of Democratic governor John Bel Edwards.[69] Having cofounded LFF in 1997 with Tony Perkins, a Louisiana native and former state legislator who now heads the Family Research Council, Mills is one of the state's most influential Republican operatives.[70] He is also a "Seven Mountains" dominionist who has openly advocated the takeover of all cultural and governmental institutions by evangelical Christians (that is, those who share his particular views).[71]

Seven Mountains dominionism, according to Clarkson, "calls for believers to take control over seven leading aspects of culture: family, religion, education, media, entertainment, business, and government. The name is derived from the biblical book of Isaiah 2:2 (New King James Version): 'Now it shall come to pass in the latter days that the mountain of the Lord's house shall be established on the top of the mountains.'"[72] Mills publicly prayed for such a takeover at a January 2015 prayer rally held in a sports assembly center at Louisiana State University. Warning the audience that "these seven spheres of influence are under enemy occupation right now," Mills appealed directly to God: "Father, we cry out for the seven mountains of influence today. We pray that you will give us government, arts and entertainment, education, the church, and the family. That our ambassadors would occupy the high places. That you would bring us into a place of understanding that they need to be occupied by the body of Christ because it's rightfully His."[73] Jindal personally helped organize this event at a prayer meeting at the Louisiana governor's mansion.[74]

Christian evangelicals' political influence in the United States has been achieved predominantly through their alliance with the Republican Party, for which they are an essential source of support.[75] That is true in Louisiana as well. However, Jindal and Mills were aided in their dominionist efforts by the fact that they met with little (and sometimes no) opposition from the Louisiana Democratic Party. In Louisiana, party lines tend to break down where religion is concerned. Many (but certainly not all) Democratic legislators share the LFF's anti-secularism and are willing allies. Prominent Louisiana political columnist Jeremy Alford has noted that Mills has personally, "in his own name" rather than on behalf of LFF, endorsed candidates and even donated money to the campaigns of conservative Democrats, upon whom he can usually count for legislative support; in this way, says Alford, LFF "gets involved in the election process without overstepping legal boundaries."[76] Former Democratic state senator Ben Nevers, who sponsored the LSEA on LFF's behalf and was one of the state's most powerful elected officials until term limits ended his senatorial career, was a firm ally of LFF on matters related to the Religious Right agenda.[77] Although Republicans were the majority in both houses of the Louisiana legislature during Jindal's second term (2012–2016), both the 2008 LSEA and the 2009 human-hybrid bill were passed when Democrats were in the majority during his first term (2008–2012).[78]

Such bipartisan support, cultivated through his virtually constant lobbying, has amplified Mills's influence far beyond LFF's small institutional size. Moreover, LFF's statewide network of ever-ready auxiliaries—pastors and churches—is always ready to spring into action. Mills has stated (diplomatically, given the tax-exempt IRS status of these auxiliaries) that "there are chaplains that have begun to circulate in the Capitol during the legislative session, to pray and to minister to the needs—not to lobby, not to provide policy input, but simply to be a present help for those who are on call, for those who are on duty."[79] LFF also holds an annual "Legislative Pastors' Briefing" immediately before every legislative session, with elected officials in attendance.[80] Afterward, LFF holds an annual "Legislative Awards Gala" to recognize legislators who vote favorably on its issues.[81] Prior to every election for state offices, LFF publishes voter guides that indicate where candidates stand on issues related to its agenda, followed by widely publicized legislative "scorecards" after every legislative session.[82] Consequently, legislators fear Mills's opposition to their bills, and some politicians, both those already elected and those running for office, purchase sponsorships for the legislative galas. For example, Mills has a close, long-time association with current Louisiana attorney general Jeff Landry, who has been a financial sponsor of the gala and partners with LFF in its anti-LGBT activities.[83] Louisiana political science professor Pearson Cross accurately describes LFF's role in Louisiana politics:

As the most powerful non-business lobby in Louisiana, LFF has demonstrated a remarkable ability to sway, persuade, cajole, threaten and encourage legislators to support its agenda.

Its agenda, in turn, typically focuses on tightening regulation on abortion providers, challenging the rights of the LGBT community, channeling public money to private schools, asserting the primacy of Christianity over all other religions and demonizing opponents as Satan-inspired.

Led by Gene Mills, a former youth pastor with close ties to Tony Perkins of the Family Research Council and former Gov. Bobby Jindal, the LFF has succeeded in Louisiana for three reasons: it has pushed aside other challengers and established itself as the loudest and most persistent voice of faith and family in Louisiana; it has built firm ties with conservative evangelical churches all across Louisiana; and it has immersed itself in legislative politics.

The success that LFF has had in legislative politics depends on the close relationships formed by leader Mills with legislators, the large number of deeply religious and conservative legislators elected in Louisiana and the success of events like ... [the] Legislative Awards Gala, which ... [celebrates] legislators who supported or "voted with" the LFF 100 percent of the time, 90 percent of the time and so on. The vast majority of these legislators are Republicans, although a few Democrats ... top the LFF's list regularly.[84]

LFF also cultivates relationships with United States congressmen and senators. For example, Louisiana congressman Steve Scalise, who was severely wounded in a shooting in 2017, was the recipient of LFF's highest award, the Gladiator Award, at its 2017 legislative gala; Scalise is in philosophical lockstep with LFF.[85] In keeping with the Religious Right's emphasis on shaping the federal judiciary, LFF actively (and successfully) promoted Louisiana attorney Kyle Duncan, another Gladiator Award recipient, whom state attorney general Landry calls "the Neil Gorsuch of Louisiana," as a nominee to the Fifth Circuit Court of Appeals.[86] Donald Trump obliged by nominating Duncan, who was narrowly confirmed by the Republican majority in a party-line vote in the United States Senate.[87] Mills is a vocal admirer of Trump, despite the latter's moral failings in

the areas of faith and family that Mills touts as the foundation of government. After attending the 2018 National Prayer Breakfast, Mills quoted
Trump approvingly concerning the supernatural foundation of our rights
as Americans:

> There, the President correctly reflected on the origins of "our rights,"
> Trump said, which "are not given to us by man. Our rights come from
> our Creator. . . . No earthly force can take those rights away." Then, in
> an affirmation of religious liberty, he echoed, "When Americans are
> able to live by their convictions, to speak openly of their faith, and to
> teach their children what is right, our families thrive, our communities
> flourish, and our nation can achieve anything at all." To these senti-
> ments, LFF says Amen![88]

IN THEIR OWN WORDS: HOW TO ACHIEVE DOMINION IN A DEMOCRACY

Although the perspectives of scholarly observers such as Cross are valuable, Jindal and Mills provided revealing insights into their own views of
their roles in de-secularizing state government during Jindal's governorship. Mills frequently reminded supporters (and prospective donors) that
LFF was holding "gatherings like the monthly pastors fellowship lun-
cheon with Louisiana's Gov. Bobby Jindal at the Governor's Mansion to
break bread, share the word and pray together!"[89] While these religious-
political networking events were private, others, such as LFF's December
2007 "Governors' Gala" celebrating Jindal's election, were quite public.
In the presence of hundreds of citizens, clergy (to whom Mills referred as
LFF's "faith arena"), and elected officials (including several former governors), Jindal sermonized that Louisianans must build up the state, "motivated by the Holy Spirit, inspired by *His* written word, in faith that He
is King and Almighty Ruler"; with Mills presiding, he also underwent
the ceremonial laying on of hands.[90] Jindal frequently—and publicly—
underwent this ritual, which pastors viewed not only as their blessing of
him but also as their transfer of God's spirit through themselves to him
and the governorship. His participation signified his submission to the

pastors' expectation that he would use his elected office to implement God's (i.e., their) will.[91]

LFF's mission is "to persuasively present biblical principles in the centers of influence on issues affecting the family through research, communication, and networking."[92] For LFF, the center of influence is the Louisiana state capitol in Baton Rouge. In keeping with the status that LFF and its supporters conferred upon him, Jindal issued executive prayer proclamations that Mills used as evidence of LFF's influence (with both the governor and God) and, most important, signed the bills that Mills successfully steered through the legislative process.[93] New Orleans journalist Clancy DuBos, listing LFF among the "big winners" in the 2012 legislative session (after only Jindal himself and the oil and gas industry), wrote that "some lawmakers reportedly were summoned to meet with LFF leader Gene Mills on the 4th floor of the Capitol—home of the governor's office—leading to speculation that LFF is now an extension of Team Jindal . . . or is it the other way around?"[94]

Mills is quite clearheaded about the power that he wielded in the Jindal administration and about his continuing role as a Religious Right lobbyist. In a 2012 article that he wrote for an evangelical Christian public relations magazine, Mills describes himself as a "missionary to the field of government," executing LFF's mission of rendering "assistance to the local pastor and the body of Christ in order that they may discover their jurisdictional authority in the arena of government."[95] He works through a "cooperating network of 300 to 500 churches [that] affords me the opportunity to have an influential voice in addressing cultural corrections and to execute objectives in an arena which respects power but often confuses the churches' authority."[96] Yet, stressing that "the vision [for Louisiana] is not complete," he says that LFF must "find, train and retain laborers to serve in this fertile mission field," ensuring that a new generation will learn "God's idea of government (jurisdiction, law, representation, continuity and authority)."[97]

Mills's understanding of the separation of church and state is, like that of ECT and the *MD*, quite one-sided; in his view, the First Amendment means that government must keep its hands off the church, but the church is under no reciprocal obligation to refrain from trying to influence the laws and policies that affect people outside the church. After the 2014 legislative session, Louisiana reporter Sue Lincoln questioned Mills about how LFF's mission squares with the wall of separation between church and state. His response was predictably self-serving:

> I asked Mills how [LFF's mission] synchronizes with the wall of separation of church and state. "In the original sense of the separation of church and state, the government has very little business—if any at all—with the affairs of the church," Mills stated. "That wall was designed to prevent the government from intruding on faith and family institutions." On the other hand, he says church influence on government is completely acceptable. "It's perfectly constitutional, and we think it's healthy," he said.[98]

Jindal's and Mills's words and deeds provide substantive evidence that they were working consciously to transform Louisiana into a de facto general theocracy. Moreover, everything that he and Jindal accomplished was done using the mechanisms of democratic government. Jindal's governorship is now over; however, the damage that these two men did remains, and Mills's work continues. The LSEA and the human-animal hybrid law remain on the books and will be there indefinitely. Fortunately, LFF recently lost its only anti-science battle since Jindal's departure. In 2017, working through a supporter who managed to get appointed to the Louisiana Department of Education's Standards Review Committee, which oversaw the revision of the state science standards for public schools, LFF tried and failed to water down the standards' coverage of the teaching of evolution.[99] Such episodes highlight the need for continued vigilance. In the meantime, Mills bides his time waiting for the election of the next Republican governor, all the while continuing his cultivation of state and national legislators and other influential figures, as he has done for the past two decades.

MADISON'S WARNING

Louisiana has exemplified some of the worst infractions against both science and democracy that the Founders tried to prevent: a Catholic, Republican governor colluded with a Pentecostal, Republican preacher, buttressed by politically organized clergy and compliant (or perhaps opportunistic) legislators, to codify their religious beliefs as law. Moreover, this was done with little opposition from either the state's Democrats or its scientists (with a few notable exceptions). The latter fact would surely be as distressing to the Founders as the transgressions themselves. Madison's "Detached Memoranda," written sometime after he left office in 1817, reveal that he could already see a similar apathy developing. Lamenting that "silent accumulations & encroachments by Ecclesiastical Bodies have not sufficiently engaged attention in the US," he warned against the "danger of a direct mixture of Religion & civil Government."[100]

The Founders bequeathed to us a secular democracy in which politicians viewed science as essential to national well-being. Yet, despite enjoying the benefits of both, Republican politicians today, as political scholars Thomas Mann and Norman Ornstein have forthrightly pointed out, have exchanged the expansive vision of the Founders for dogmatic political expediency: "The GOP has become an insurgent outlier in American politics. It is ideologically extreme ... [and] unmoved by conventional understanding of facts, evidence and science."[101] They have betrayed the Founders' legacy by co-opting the democratic system in order to help execute the Religious Right's attack on science.

PART III

PERVERTED SCIENCE, DISFIGURED DEMOCRACY

CHAPTER 8

THE RETURN OF DETERMINISM: SCIENCE, POWER, AND SIRENS IN DISTRESS

KURT JACOBSEN AND ALBA ALEXANDER

Rational, adj. Devoid of all delusions save those of observation, experience, and reflection.
—Ambrose Bierce, *The Devil's Dictionary*

What Ambrose Bierce in his usual fierce deadpan manner meant is that observation, experience, reflection, and, we daresay, logic do not necessarily save us from delusions, as anyone with more than a skin-deep understanding of the history and philosophy of science will appreciate.[1] Freeze or idealize any method and, presto, you have a catechism that rules out valid and valuable avenues of inquiry. Positivism, anyone? Skinnerian behaviorism? How about rational choice? All these secular priesthoods, as anyone who has crossed swords with them will attest, would like nothing better than to excommunicate anyone they deem heretics.

Scientism is the use of theory to explain reality, as one unknown wag concisely put it, while science is the use of reality to explain theory. Pushed beyond their proper domain, and with all caveats shaken off, these methods instead become barriers to investigative thought.[2] "Science and religion have changed places," philosopher Slavoj Žižek, who can always be counted on for hyperbolic accuracy, observes. "Today science provides the security religion once guaranteed."[3] John Gray adds, "For us, science is a refuge from uncertainties, promising—and in some measure delivering—the miracle of freedom from thought, while churches have become sanctuaries for doubt."[4] The ironies are acute since organized religion was the original culprit in radical diagnoses of alienation.

A pervasive hitch in jeremiads such as Chris Mooney's *The Republican War on Science* is that they attribute repugnant decisions by the George W. Bush or the Trump administration today to sheer ignorance of or else a mere misconstruing of science, rather than to the perpetrators' interest-driven agendas to which they force all else to conform. One need not dally much with Machiavelli to realize that it is customary for political factions to pounce on absolutely all events in order to interpret them to their advantage.[5] Imagining that political figures care foremost about confining themselves to the dictates of science (or even of the "intelligence community") is pretty much the Olympian height of naiveté.[6] Expecting rationality to reign at these conniving high levels is, one might say, quite irrational. Demanding rationality is fine and necessary, but what exactly is it that is demanded?

The Republicans, in their collective continuous King Canute conniption fit, oppose global climate-change arguments, embryonic stem research, and evolution. Trump, on his sordid campaign trail, promised to dump the EPA withdraw from the Paris climate accord, and indeed repudiated the very idea of climate change.[7] Following the money, not scripture, is much more revealing of bedrock motives. Republican atavism is rooted less in medieval lore or tribal fancies than in pecuniary concerns of corporate players who enlist any credulous ally in the fight against restrictions on their quest to commodify everything in sight for power and profit.[8] From their lofty Wall Street point of view it is always the perfectly rational thing to do. Nothing can be dispensed properly without shelling out a profit for it to some chokepoint supplier. Without serious megadonors like the Kochs and their clout, the revived Scopes trial image of modern America, allegedly overrun by petit bourgeois Tea Partiers, would recede into marginalized backwater grousing where it belongs.[9]

So irrationality isn't evidenced only in braying that the government should keep its paws off social security or insisting that humans coexisted

with velociraptors or reading horoscopes or imagining Elvis lives. Irrationality includes believing tax cuts make everybody better off, that austerity
policies cure economic stagnation, that high-tech military coercion is
wise foreign policy, that Saddam Hussein had nukes ready to fly in forty-
five minutes flat, and that indecipherable financial derivative instruments
make markets safe and sound.[10] One finds plenty of reputable folks who
espoused or espouse all those extremely dicey propositions. Some even
accrue Nobel prizes and cabinet appointments. Distinctions between
delusions and blatant lies might be made in all the preceding cases, but we
usually cannot know for sure which is which. Bleeding, after all, once was
the last word in medical practice; phrenology and eugenics were respectable and even revered sciences. So we're dealing with a continuum of irrationality wherever we tread.

The insidious predicament is that irrationality dons a host of pleasing
shapes, including, most effective of all, the guise of bright shining
rationality itself.[11] This is hardly a surprise. Karl Mannheim (and Max
Weber) had good reason to divide forms of rationality between instrumental ("subordination of all means to a single end") and substantive
("insight into complex interrelations of matter"), because the former
version encourages unchecked excess and ultimately mad actions. From
this source we eventually get the paranoid certainties of closed system
thinking;[12] and the remorseless narrow centralizing view of what James
C. Scott calls "seeing like a state"—which is applicable to any powerful
entity, public or private.[13]

No enterprise, not even science, is immune to manipulation, to being
bent to the purposes of the wily agent. The satanic figure in ecclesiastical
garb haunts all religions, with Spanish inquisitions, Witchfinder generals,
and Dostoevsky's not entirely fictional grand inquisitor sowing horrors
in the quest to stamp out heretical irrationalities and rationalities alike.
Determinism is just the secularization of church dogma, the subordination of intellect to a single narrow track of explanation for any phenomenon. "The mechanical philosophy of Robert Boyle was integral with his
natural theology," Robert M. Young notes:[14]

Newton's mathematical principles were of a piece with his Biblical ruminations. Charles Lyell's uniformitarian geology was carefully composed so as to be consistent with his views on Genesis, and he never deviated from believing in the separate creation of man, no matter how much he conceded to Darwin with respect to the evolution of other forms of life. Similar claims ... were later to be convincingly made about Darwin.[15]

Early nineteenth century French political theorist Henri de Saint-Simon, who apparently knew something about human psychology, started Newtonian temples based on allegedly scientific principles, which were "never simply contemplative, but consciously manipulative in design."[16] Charles Sanders Peirce later anticipated, in the purest pragmatic faith, arrival at the single "opinion that is fated to be ultimately agreed to by all" through scientific means, which is what scholars actually intend whenever on a futile quest to establish a unitary social science or a unitary discipline of any kind.[17] The dream of all demagogic intellects is to force all thought, all imagination, all inquiry down a common route, congenial to their own emotional needs and institutional aims.

In economics, neoliberals want to liberate us through the tender mercies of the unchecked but state-aided market, which permits no other gods before it. Positivists want to free intellect by chaining it to their desiccated creed, which permits no examination of the social relations in which they themselves are reverently embedded.[18] Science for them must have one format or else it just isn't science.[19] Hermeneutics, phenomenology, and psychoanalysis do not count; they *literally* do not count in this perpetual haste to worship the mystique of numbers. Who then legitimately wears the mantle of rationality? "In fear of the growth of an irrational anti-science movement, researchers and those who yearn for the benefits of scientific research have doubled their efforts to dampen the effects of criticism," Wendell Wallach recently noted.[20] "In other words, even criticism contributed to driving forward the technological juggernaut." That is our counterintuitive theme here.

REASON VERSUS RATIONALITY

Rationality can be as much a crass self-serving slogan as it is a term comprising a desirable means of handling reality; the capturing of the emotive term—and it is emotive—as a standard for one's paradigmatic preferences is one of the key political moves used every day in the social sciences in battles between different branches and schools. The reason why Max Horkheimer and the Frankfurt School defined rationality as incorporating an emancipatory intent is because a one-dimensional form of rationality leads not only to dead ends but dreadful ones too.[21] Absent a conscious emancipatory intent to ground rationality in a dynamic critical view of social relations, science easily shifts into and is subordinated to elite agendas of domination. Rationality, when drawing unnecessary inflexible borders around admissible knowledge, detracts from reason.

The resistible rise of rational choice, especially in political science and sociology, is a case of primal urges—the desperate seeking of certainty—posing as saintly bearers of unalloyed truth, whose followers imagine everyone else is less "scientific" than they are. Does the fellow in the white lab coat have anything in common with the yahoo beneath white sheets and hood? They most certainly had crucial things in common during the heyday of eugenics when they shared a crusader's sense of serving the one true faith. These white-garbed warriors backed the Ahabian motto of all the means being sane but the goal being mad. In this same vein Jacques Ellul later dryly summed up the onset of "the technological society" as the increasing application of precise means to carelessly examined ends.[22] Yet just a few decades ago this state of affairs had been severely challenged to the point of near extinction.

"Once the restrictive canons of positivism and the naive belief in objectivity were discredited," Barry Richards wrote in the 1980s, "the reasoning for ignoring psychoanalysis or phenomenology or anything else on the grounds that it was not objective science—was fatally weakened."[23] True at the time. Unfortunately, they came roaring back because the underlying thrust, the emotional and the institutional craving for easy or complex certainties, propelled them back to prominence again.

Genetic perfection (despite eugenics), "high tech war-winning weapons" (despite Vietnam), and wonder drugs (despite the anti-psychiatry movement) were far too attractive as pseudo-explanatory tropes to stay down for long.

By the 1970s technological determinism, as an ubiquitous cultural fixation, had lost much of the persuasive power it enjoyed in the early postwar era. Critiques by Lewis Mumford, David Noble, Reinhard Skinner, Langdon Winner, the Frankfurt School, and other eminent analysts subdued the naive zeal animating giddy visions of cure-all gadgetry.[24] Yet technological determinism was only pushed shyly to the side, where it was remained the creed of choice by technically proficient practitioners at places like MIT, the computer and telecommunications industries, or the wider encompassing military-industrial complex, which see no reason to quibble at the determinative power of the instruments they wield, especially as those instruments underwrite their paychecks.[25]

"What counts is what can be counted," is a Laingian double-bind phrase that pretends to liberate inquiry at the same time as it crimps it.[26] The slogan, paraphrased, is emblazoned above the Social Science Research Building at the University of Chicago. And it's crazily restrictive, the repository of the lazily industrious mind, all too pleased to settle on a socially approved proposition. The allure of premature certainty beckons us through this widespread Philosophy 101 version of rational activity, an activity, as Hannah Arendt admonished, that was ultimately dehumanizing, as we observe what once were fellow creatures but now are gnats from a distant Archimedean vantage point.[27]

One unhappy irony of the struggle against Republican Party intransigence on climate change, and similar idiocies, is that supposedly pro-science champions skip merrily past all the withering critiques that had worked so well to keep determinist fancies at bay. A restored certitude arises among ardent defenders of science that rivals any televangelical con man working the crowd or airwaves. Science becomes pretty much what clueless positivists say it is and nothing more. The scene begins to take on the character of a B-movie where a scientist assures everyone that there is a perfectly rational explanation just as the monster descends. One is supposed to forget that behind scientific authorities is a panoply of institutional interests they work for in a commercial realm of ever-more maldistributed power.[28] Determinist ideologies of this sort suit authorities because they are peddled and packaged by the same authorities to shore themselves up.

So determinism, in many beguiling forms, slips back into public discourse. We now are supposed to know that there is a pinpointable gene for every behavioral quirk, a rising from the grave of eugenics in spiffy new embodiment. We supposedly know in the same dubious way that pharmaceuticals "correct" chemical imbalances and dopamine deficiencies. Half the United States now is drugged to the gills, mostly legally. Finally we trust that statistical apparatuses are utterly reliable guides to military security matters. Underlying all these confident claims is a resurgent faith in the old time religion of obsolete nineteenth-century notions of science. Against the allure of those notions Wallach cites the classic but deeply underappreciated conundrum that each decrease in "black swans" just "adds layers of complexity that breed more black swans," which as Schwartz long ago noted leads to underestimates of occurrence of "low-probability" misfortunes," and then "to a chain reaction of unanticipated events or underanticipated events."[29] Only a dynamic sensibility about science can cope with or avoid the primly static paths that exacerbate these problems.

In Vietnam, the Hamlet Evaluation System (HES) and the entire contrived menu of McNamara's cherished indicators attested to a progress that, under scrutiny, turned out to be a figment of the measures themselves. In economics, derivatives were hailed as wise instruments for market protection until they caved in the economy. The economists were and are bedazzled by the supra-reality of numbers, as if numbers themselves were not metaphorical. Defenders of science often embrace overhyped depictions of the causal magic of genes, of brain chemistry, of artificial intelligence, and of magic bullet warfare. These latter phenomena are modern siren songs, not irrefutable arguments, and as usual they obscure underlying key social forces and purposes at work.

A emblematic 2017 London event entitled "A Celebration of Science

& Reason"—and a mere snip at 117 pounds a ticket (who can afford their brand of reason?)—featured Richard Dawkins and Sam Harris, two chaps as bound by blinkered views of science as any Bible belter is by their own vaunted strict interpretation. The counterattack on much plainer fools makes any professed enemy of superstition and tradition appear to be a gallant enlightenment hero. Scientism, a misapplication of a form of science from one domain to another in which it is not suited, is a persistent pestering presence. The best scientists know their techniques are not universal skeleton keys, but not all scientists are the best scientists as any scan of scientific scandals will attest.[30]

Many scholars, apart from the Frankfurt School, warned against overvaluing dull instrumental logical apparatuses for unraveling tough human puzzles. The author of *The Republican War on Science* blithely regards anyone who disputes the soundness of genetic engineering as an idiot, and invokes Steven Pinker to grouse that "the genetic underpinnings of human behavior" have often gone unstudied because of a "general left-of-centre sensibility that anything to do with genes is bad."[31] Pinker is enamored of the "notion that a tendency toward violence might have genetic roots" even or especially if it raises among the spoilsport Left "the fear of Eugenic-style solutions."[32] This latter accusation conveniently is scored as an "abuse of science" for which conservatives award themselves points. Yet the fundamental question underlying these debates is, "Are we containers of objectified forces," as Smail succinctly puts it, "or do we pursue our own needs and wants?"[33]

GENETICS: REGRESSION TO THE MEANEST?

Science and medicine in the nineteenth and early twentieth centuries, saturated by a casually racist and class-ridden Western culture, generated the grim crusade of eugenics, and there is nothing in twenty-first-century life to lead us to expect that genetic research, even if conducted by improbably angelic sages, will be free of originating eugenic ambitions. Even the British Medical Association soberly had to admit that despite "efforts constantly made to distinguish between eugenics and genetics, the two are clearly related."[34] The best stock, you see, must win out, although many fretful upper-class specimens concluded that they had to fight for their survival against a rising tide of mediocre biological material constituting the masses. The Nazis took this stern frame of mind to the limit and beyond, though Nazism was, according to one pithy summation, "nothing but applied biology."[35] A happy eugenic ending requires a bit of ruthless pruning.

The seductive notion of good stock—always "us"—remains quietly pervasive in popular culture. In Western industrial countries this terribly self-flattering idea hooked up with technological supremacy, which was believed to accompany through some sort of divine providence one's superior biological heritage. For a "generation fearful of race decline, the new technology seemed to buttress the dominant position of the white race." Michael Sherry noted of the turbulent interwar era, "The vision of air war that emerged before WW II was shaped by the broad currents of racism and faith in progress."[36] How lucky affluent WASPs were that, as Max Weber reminds us their clergy preached, virtue and power strutted hand in hand.[37] The "nullity" of "specialists without spirit, sensualists without heart," as Weber foretold, now run rife in assured urbane pedigreed ways. It is, happily for the winners, genetic destiny.

Late nineteenth-century race theory imposed, in Mahmood Mamdani's words, a "marker dividing humanity into a few superhuman and the rest less than human, the former civilized, the latter putty for a civilizational project."[38] The harrowing Nazi experience was the crest point for eugenics, but the dotty dream of biological perfection—perfect relative to what?—persisted, and eugenic sterilization went on being practiced in some US states well into the 1970s, targeting the poor and minorities. In the tidy eugenic universe, if you lack resources to fight back you clearly count as a confirming instance of inferiority. Eugenics did not so much disappear after 1945 as go stealthy. The animating conceits clung tenaciously on.

A core conceptual conceit here is the notion that genes are perfectly self-contained single action entities designed to produce invariant outcomes: blue eyes, snub noses, schizophrenia, and inclinations to obey or break traffic laws. Yet the most reputable investigators discovered regarding genes that, far from being single-action or immune to environment, "redundancy seems to be built into gene functions, where, if one is 'knocked out' others seem to supply the same function," Evelyn Fox Keller summarizes, which ought to pose an insuperable problem for the popularized "explanatory framework of the genetic paradigm."[39] Nonetheless, the newspaper and media still overflow with breathless routine announcements of discoveries of genes for this or that or the other behavioral trait.

Negative eugenics augured mass murder; positive eugenics involves prenatal genetic testing, in vitro screening, and prenatal testing for the purpose of preventing "carriers" of undesired genes. The laudable aim of curing diseases is deployed to camouflage the underlying and undying elite dream of exerting complete control over the behavior and the beliefs of the population.[40] A generously funded apparatus of labs, and an accompanying medley of credulous media outlets, have been tilting the nature versus nurture debate firmly toward nature. Yet the Human Genome Project, ballyhooed to the heavens, came up with distressingly paltry results, given what they expected to find.[41]

The first schizophrenia gene allegedly was nabbed at University College London in 1991, but nobody since has replicated this or any other such claim of genes for mental maladies.[42] "The hypothesis that enduring psychosis/schizophrenia has strong genetic underpinnings," a 2016 survey concludes, "is not supported by available evidence."[43] David Plomin remained hopeful, though puzzled, as to why genetic research into behavioral traits has yielded so little.[44] The next backstop was to resort to the proposition that it is likely that there are "a number of susceptibility genes which interact with one another and with environmental effects."[45] That extenuating collective tack did not work out either, but implacable pursuit squads of researchers remain sure that the right amount of money, technical devices, and pluck ultimately will snare the long-sought genetic culprits galloping away just over that next high ridge.

What then is the statistical significance outcome of all the claims that a gene, or a distinct group of genes, has been located with exactitude as causing inclinations to be gay or to bite your nails or to question authority? Well, it's around zero point zero something. On that patently flimsy basis one daily encounters scientists—securely cocooned in upper-middle-class lives—boldly stating that we must not be afraid to "face facts," just as German scientists in the 1930s faced facts of their own making and did what was required. One need not have visited Nazi Germany though, which took its cue from Indiana and North Carolina sterilizations prescribed for lower orders, especially if not white enough. Let us suppose that hardnosed scientific authorities like Charles Murray or James Watson found that most unexpected of findings: that their own genetic composition was inferior or "tainted" (as eugenicists liked to put it). Would they then have agreed blithely to have their bloodlines rubbed out? These are the baleful blinkered breed of scientists that Kurt Vonnegut, putting his University of Chicago anthropology experience to some good use, mercilessly satirized in *Cat's Cradle*, and they are always around because class society is always around and neuroses too are always with us.

The cumulative result in the research sweepstakes for larger portions of funds for diminishing results is significance ratings approaching the odds of drawing a royal flush on first deal. Jay Joseph and Claudia Chaufan piquantly note a Colorado Adoption Project twins study that found "the mean personality scale correlation between birthparents and their adopted-away biological offspring"—a relationship that they considered "the most powerful adoption design for estimating genetic influence"—was zero."[46]

The meaning of heritability is conveniently confusing so as to mislead the public to believe that there is a direct percentage of probability for the occurrence of each disease or trait. Joseph and Chaufan, though, point out the elementary fact "that genes and environments do not build phenotypes independently, so there is no common unit of measurement that enables us to say that in one individual, genes "caused" x percent of some trait, while the environment caused y percent."[47] The press, culturally gullible and opportunist, chooses to print the myths. The real loss is to studies of environmental influence in mental health maladies.[48] According to the dogma, no parent, or set of social conditions, anywhere at any time can ever inflict mental damage on a child who is not predisposed to suffer it anyway. One can see why parental organizations adore this congenial line of argument.

What about Pinker's whining about left-wing interference? "In 1965 a team of serious researchers suggested that XYY males were more aggressive," Alvin Rosenfeld noted. "That finding was disseminated widely and contributed to a heated debate about behavioral genetics. In 1993 a report for the National Academy of Science dismissed the link as unproven and nothing of the kind since has withstood scrutiny. In those intervening years, many parents were warned about their XYY children's potential for aggressiveness." How many families were hurt, he asks, "because parents were worried that their children might grow up to be murderers?"[49] Pinker, however, is dead sure that science will find otherwise. As David Bell aptly writes of Pinker's latest tome *Enlightenment Now*, his stance is one "in which data and code are all too often held to trump serious critical reasoning and the wealth of the humanistic tradition and of morally driven activism is dismissed in favor of supposedly impartial scientific and technological expertise."[50] Just keep pumping money and you're sure to find something that pleases Pinker. Or are you?

Ruth Hubbard reports that DNA neither "self-replicates" nor "self-transcribes" nor "self-translates": the cell, in response to a host of internal and external signals, "transcribes segments of DNA and translates the resulting RNAs into proteins."[51] The notion that environmentally induced changes can be inherited has long been regarded as Lamarckian folly, but as Stuart Newman points out, it is now being incorporated with whatever degree of chagrin into evolutionary theory.[52] As David Moore writes, "We now know that DNA cannot be thought of as containing a specific code that specifies particular predetermined (or context independent) outcomes."[53] A NASA study recently found that 7 percent of the DNA of an astronaut in space for a year altered permanently from the experience.[54]

In Evelyn Fox Keller's words, DNA "does not even encode a program for development."[55] The idea of a predetermined code, of DNA as the rigid template of heredity in which our fates are transcribed, is at odds with phenotypic plasticity, the ability of organisms to adapt to the demands of their particular ecological niches." What is it to say, asks Evan Charney, "that all the features of a person are in her genes, but to take all the attributes of a person and ascribe them (or their "predispositions") wholesale to what is in effect a little latent person—the inherited homunculus-genome? The reigning ideology of DNA."[56] Indeed.

None of this is news to anyone paying attention outside mission-oriented research bubbles of gene hunters.[57] Psychiatrist Hervey Cleckley in 1941 could observe presciently that "distinctions between organic and psychogenic are sometimes far from absolute."[58] The organism changes in response to every item of experience. "It would not be profitable to confine our concept of what is organic to the cellular level with so much already known which indicates that molecular and submolecular changes (colloidal, electro-chemical, et) are regularly resulting from our acts of learning, or, if one prefers, from all of our conditioning."

One can cite many more examples of clear-eyed misgivings about gene research, all the way back to the 1940s and earlier. All were ignored by researchers avid to prove otherwise. This current credulous phenomenon promises a farcical repeat of the eugenics reign, where a dubious but popular creed moved side by side with Mendelian scientists whose investigations refuted the whole basis on which eugenics enthroned itself. The word somehow did not get out. Ignoring refutations is the conventional researcher's first line of defense. Philosopher of science Thomas Kuhn, according to some of interpreters, seemingly approved of this conservative methodological practice as one that assured that only the hardiest anomalies would rupture a reigning paradigm.[59] Jonathan Leo recalls,

> While there are famous adoption studies and there are famous twin studies, these are separate studies. The famous adoption studies did not study adopted away twins. I once heard the Department Chair of a major university Psychiatry Department make the same mistake in a seminar, and when challenged about it he, and the entire room of biological psychiatrists, stood by the claim. Be careful of the echo chamber in your field.[60]

Every scientific field, as inevitable in any institution, features its echo chambers, sometimes known as "epistemological communities."[61] Physicians Michael Joyner, Nigel Paneth, and John Ioannidis found that $15 billion of the $26 billion NIH extramural funding was linked to the search terms: gene, genome, stem cells, or regenerative medicine—but yielded little in worthwhile results.[62] One wonders if several decades hence this enterprise will be viewed as a Western-style Lysenkoist folly, where the weight of both corporate and enlisted state power expensively sustain a dead-end research program because it tickled the egoistic fancies and social agendas of the funders.[63]

BIG PHARMA AS PANACEA

Genetic determinism ties in all too symbiotically with pharmacological determinism. The familiar kindred credo is that one miracle pill makes you bigger and a different pill makes you smaller. The public is saturated with canny adverts—trailing interminable strings of small print caveats about side effects—boasting that there is just the right drug available to elevate troubled people from or plunge them into any designated mood or physical condition.[64] If you suffer from a personality disorder, it is because of a serotonin imbalance, or if you are depressed it is because of a lack of norepinephrine. Rather like a basket of limes cure a ship full of scurvy.[65]

The pharmaceutical industry wants everyone to understand that their patent medicines—like genetic determinists claim for genes—exert precise one-to-one actions targeting our individual ills, woes, and even existential insecurities. Cyanide or arsenic in doses adequate to the fatal purpose may display such properties, but few other drugs do. Opioids apparently work all too well in their limited domain of relieving pain, to the point of snuffing out 64,000 Americans (especially in Appalachia and the Southwest) by overdose in 2016 and being declared a public health emergency in 2017 (without needed funding to address it).[66]

In 2003, to curiously scant media fanfare, a senior executive with

Britain's biggest drug firm admitted that most prescription medicines do not work on most people. Fewer than half of patients prescribed drugs seemed to benefit, however one defined "benefit."[67] In Britain his comments came shortly after it was learned that the NHS drugs bill had risen 50 percent in three years. Dr. Allen Roses, an academic geneticist from Duke University, cited figures showing that Alzheimer's disease drugs work in fewer than one in three patients, while those for cancer were effective in a quarter of patients. Drugs for migraines, osteoporosis, and arthritis work in half the patients. By the industry's own measures: drug efficacy rate in percentages was: Alzheimer's: 30, Analgesics (Cox-2): 80, Asthma: 60, Cardiac Arrhythmias: 60, Depression (SSRI): 62, Diabetes: 57, Hepatitis C (HCV): 47, Incontinence: 40, Migraine (acute): 52, Migraine (prophylaxis): 50, Oncology: 25, Rheumatoid arthritis: 50, Schizophrenia: 60.[68] Side effects don't even come into it. Nor, since the start of the millennium, does it seem to matter that 100,000 die a year from adverse drug reactions and a million or more are harmed by drug reactions that require hospitalization.[69] For psychiatric drugs "the latest best estimates as to the percentage of people who benefit over and above placebo effects are 20% for antipsychotics and even less for antidepressants."[70] John Read, Olga Runciman, and Jacqui Dillon reported that

> A survey of 1829 people on antidepressants found the following rates: sexual difficulties (62%), feeling emotionally numb (60%), withdrawal effects (55%), feeling not like myself (52%), agitation (47%), reduction in positive feelings (42%), caring less about others (39%), and suicidality (39%).

They also note that in twenty-four of twenty-five nations surveyed the public believes that social factors play a bigger role than genes or chemical imbalances in causing mental health problems, with the sole exception being the United States, which speaks to the incessant institutional drumbeat message that genes and chemical imbalances cause everything we do and can't do.[71] Dawkins's selfish gene evidently is deeply neurotic.

What is most interesting for our essay about this admirably candid event is how the researchers tried to explain it. "I wouldn't say that most drugs don't work." one university researcher opined to the press.[72] The problem is not the drugs themselves or the firm's policies but rather that the "recipients carry genes that interfere in some way with the medicine." It can't possibly be because the drugs do not work in the advertised way. Therefore, it stands to reason that the genetic structure of the patient perversely prevents some of the drugs from doing their jobs. Psychopharmagenetics, as a field, arises with a mission to fit square pills into the round mouths. Nothing much has changed regarding drug efficacy in the intervening years, but the drug prescribing only goes up. Seven out ten Americans gobble prescription drugs.[73]

At the same time the US government since Nixon has conducted an extraordinarily extensive and harmful war against illegal drugs, regardless of their actual ill effects, while enshrining the right of pharmaceutical companies to peddle useless, addictive, and suppressive drugs as sure cures for purely internal ailments (for which environmental factors do not matter) we suffer. Nothing in modern life, with its infliction of insecurities on the mass of the population, allegedly can cause any mental state tantamount to depression or schizophrenia (which seems to mean anyone the diagnosing shrink fails to relate to). That is a rock solid article of faith, and a lot of already overpriced stock values are riding on it. The purpose of the drug war has far more to do with social control of surplus population—surplus, that is, to the need of the profit system for their services. Meanwhile, pharmaceutical firms casually raid the public purse for free R and D support, tax breaks, and grants.[74]

One cannot help but remark that if there were a smidgen of actual logic to these interacting determinist streams (genetics and pharmaceuticals), US authorities would give up on incarceration as punishment for an estimated 27 million illegal partakers, for they would be deemed victims of genetic urges they cannot help.[75] Then again the authorities, by dint of their rigid punitive cast of mind, only proved they were genetically coded to do what they did. One sees how tormented and contorted these glib determinisms quickly can become.

Versus all these muddled-up deterministic frames of mind stood figures like Freud whom the "hard scientists" despise. Author Peter Gay notes that psychologists of Freud's era thought he was "not biological enough," because they ultimately believed mental disturbance was rooted in "heredity or some physical trauma." Freud seemed just a soft-headed environmentalist:

> Indeed, in 1912, Freud found it necessary to defend himself, a little irritably, against the charge that he had "denied the significance of inborn (constitutional) factors because I have stressed that of infantile impressions. Such a reproach stems from the narrowness of the causal needs—*Kausalbedürfnis*—of mankind," which likes to posit a single cause if at all possible.[76]

So Freud tartly identified the primordial urge underlying determinism and its staying (and returning) power.

MILITARY MATH

The final case of the return of irrational determinism that we will address is a military instance. For all the routine reverential comments one hears in policy circles about heeding military theorist Carl von Clausewitz's admonitions about the "fog of war," leaders, especially in superpowers, like to imagine they can enforce their will on others through the right algorithmic schemes. Take Vietnam: "[Robert] McNamara was both the product and the servant of a society that likes to express itself in the grammar of violence, and he was caught up in a dream of power that substituted the databases of a preferred fiction for the texts of common fact," Lewis Lapham writes. "What was real was the image of war that appeared on the flowcharts and computer screens. What was not real was the presence of pain, suffering, mutilation, and death."[77]

In what is really the crapshoot of war, the minders of militaries are drawn to formulaic solutions to solve vexing phenomena such as counterinsurgency operations in areas in full-scale revolt. In Vietnam, right up to and beyond the end of the war, authorities employed the statistical

Hamlet Evaluation System (HES) to concoct a sense of steady progress in extinguishing resistance in the vast countryside, which they really did largely by wiping out or rousting out the population, An 80 percent rural southern populace at the start of the US intervention was battered, pummeled, shelled, and pulverized down to 30 percent by war's end—and the other side won anyway.

The HES, surveying some twelve thousand South Vietnamese hamlets, was the numerical lynchpin in counterinsurgency science. There are of course certain queasy questions that the analysts were not allowed, let alone inclined, to ask about their codifying task. In mission-oriented science, the usual principles of science takes a coerced vacation. The ruthless criticism of everything existing is an alien concept. Still, the practitioners regard their lore as a science, a science of domination, as if the whole purpose of science to dominate man and nature. It does not pay to understand them. Humanizing the opponent can be quite inhibiting.

The Hamlet Evaluation System was reliant on reported values, not inherent values. But once enshrined in a few digits these values take on the alluring vaporous facade of credibility. There is not much distance from the blithering statistical faith that animated the Hamlet Evaluation System in Vietnam to the green screens guiding drone programs abroad today, which serve much the same function abroad and are only beginning to get public attention. To this day one encounters credulous reports that invoke the Hamlet Evaluation System survey to buttress claims that the US really won the war in Vietnam (by suppressing the Southern guerrilla insurgency) but bugged out anyway.[78] Moyar, Sorley and other "revisionists" pounce on it.[79] The only problem is that the facts the HES depicted were mostly a figment of military imagination and of field-level expeditiousness.

If you read between the lines of one US general's retrospective comments about his predicament in that Southeast Asian guerrilla war—that it "is difficult for this democracy of ours to deal with the political dimensions of insurgency" because the "arbitrary and often undemocratic controls required" do not "go down well back here at home"—one gets a sense of the chafing these people felt.[80] It's terribly hard to be a soldier in a democracy because one must strive to conceal what one needs to do. In rural Vietnam the American readily resorted to "recon by fire," torture, crop destruction, defoliation, killing of farm animals, and random killing of villagers with every military age man regarded as a traitor or instant draftee.[81]

From 1967 to 1973, HES, based on confidence scores along eighteen rankings, rated A, B, and C hamlets as "Secure," D and E were "contested." V hamlets were "enemy controlled," and therefore candidates for obliteration. Would anyone out there raise their hand to signal they deserve the lattermost ranking? If not, might you feign cooperation? "The basic objective of increasing the population living in security from the enemy was indeed achieved," according to implacable CIA chief William Colby.[82] While he claims the Mekong Delta was basically pacified by 1970, less enthusiastic observers saw a "progress" built not on converting the peasantry into South Vietnam regime fans but on erasing them altogether from the landscape. "My own research suggests that it was not a carefully crafted military strategy of counterinsurgency that led to the apparent 'pacification' of the Mekong Delta and many other areas of Vietnam by 1971," writes David Elliott, "but a policy of rural depopulation that emptied much of the countryside—probably not a tactic that should be repeated in Iraq, or even one which is relevant to the more urbanized Iraqi society."[83]

The Hamlet Evaluation System indicated a decline of 15 percent of the number of secure hamlets in February 1968. CIA analyst George Allen noted that although the "enemy had suffered heavy losses, their forces appeared to be regrouping and could mount further large-scale action in a matter of weeks."[84] Historian Gabriel Kolko at the time warned that "the question is not who claims control but who really possesses it" and that "areas, villages, and large population concentrations the NLF [National Liberation Front] operationally controls frequently cooperate in Saigon-sponsored surveys and projects to spare themselves unnecessary conflict with US and Saigon Foes"[85] None of these nagging realities were heeded up the chain of command.

In January 1972, 42.17 percent of villages supposedly harbored no NLF (Viet Cong) infrastructure, 43.86 percent experienced sporadic insurgent covert activity; only 9.9 percent suffered regular covert activity and sporadic overt activity at night. The NLF controlled 1.88 percent at night only, 2.12 percent were fully controlled by insurgents" and yet NLF forces usually maintained threatening footholds nearby.[86] Two Vietnamese generals afterward reflected on the pacification program:

> Experience indicated, however, that in due time those enemy units which had been destroyed were surfacing again. Apparently, they had been regrouped, refitted, and reorganized in base areas with the manpower and equipment infiltrated from North Vietnam. The maintenance of area security had therefore become a frustrating task, for no matter how dense our outpost system or how well motivated our troops were, the enemy could always find loopholes to penetrate and weaknesses to exploit. Ups and downs in village security were an inevitable reality we had to face.[87]

In the HES "fraud, though not rare, was less common than the understandable tendency to resolve all ambiguities in the direction the incentives for evaluation and promotion led. Gradually, it seemed, the countryside was being pleasingly pacified."[88] The operative word here is "seemed." McNamara created a means for tracking "legible progress, but also blocked a wider-ranging dialogue about what might, under these circumstances, represent progress."[89] RAND researcher Anders Sweetland in 1968 noted, "Thus far, our search for a person who feels neutral about the HES has been fruitless. People are either for it or against it, with the 'agins' outnumbering the field six to one. No measure in the theater has been so thoroughly damned."[90] RAND colleague Austin Long reevaluated Sweetland much later, gamely observing that "Sweetland found HES to be a reliable set of metrics, *if one accepted that no objective criteria for measuring pacification existed*" (italics mine).[91] This is rather like pondering, if pigs sprouted wings . . .

David Elliott contends that the boasted success of pacification in 1970–71 was "temporary and largely the result of the depopulation of large areas once controlled by the revolution, as a consequence of incessant bombing and shelling."[92] Regarding the relativity of irrationality, he found that "rational calculations are not enough, however, to explain the tenaciousness of the hard core [of the NLF]."[93] In 1971, supposedly the high-tide mark of pacification, HES conceded in 1971 that 45 percent of the officially "safe" villages were located within one hundred meters of a recent terrorist incident, and that no official was safe at night.[94] The Office of Internal Security Affairs (ISA) inside the Defense Department reported likewise that more troops were pointless and that any gains in Hamlet evaluation "were results from accounting changes" and "not from pacification progress."[95]

Historian David Hunt emphasizes the peasantry's "refusal to be pawns of modernization" and highlights that the "irony was that the people's war launched in 1959 had been defeated, but the soldiers' war, which the United States had insisted on fighting during the 1960s with massive military forces, was finally won by the enemy."[96] "In carrying out their task of making emergent guerrilla tactics legible as part of some overarching strategic vision," a recent evaluator delicately put it regarding HES, "these systems failed to approach the ontological question of what actually characterizes the supposed 'rational' or 'obedient' subject in asymmetric warfare."[97] Rationality never mattered so long as authorities imagined they could afford to be irrational in pursuit of goals that has many international relations scholars still scratching their heads. What Freud called the seductive lure of the single answer, *Kausalbedürfnis*, intersected with arrogance, which explains rather a lot about the embrace of determinist frameworks and consequent abandonment of reason.

No one ever conceded officially that the HES was fundamentally flawed in conception as well as execution. One can marvel at extraordinarily strained and tortuous appreciations of it infesting refereed academic literature today.[98] The determinist impulse behind it resurged in the war on terrorism—as has counterinsurgency theory *sans* the good sense not to apply it—and especially was exemplified with all the mumbo jumbo technology involved in drone warfare. The promise of precision guarantees victory, or so another generation of purportedly rational and highly lettered leaders imagine.[99]

Collate the data. Flip a switch. Out of the blue comes a hellfire rocket pulverizing everyone in range. This time around one supposedly wins hearts and minds by selectively, rather than indiscriminately, killing relatives and friends. The United States devises an "approach to counterinsurgency warfare and border policing," writes Gusterson, "that is organized around new strategies of information gathering, precision targeting, and reconceptualizing enemy forces as a cluster of networks and nodal leaders."[100] Collateral damage casualties are low-balled for US public consumption.[101] Afghanistan is the most heavily drone-patrolled patch of digitized land on the planet. In Iraq, drone strikes receded under Obama from sixty in Iraq in 2008 to none in 2009, and recommenced in Syria in 2014. Notice any decline in resistance as a result?

Militarily technological determinist fancies are as intoxicating as ever. These "machine dreams" resurge because their sponsors believe it is perfectly rational to deploy the latest gadgets, because they anticipate a rational domestic audience will be too busy or gulled to care, because foreign hosts will see their use as the most rational choice among a lot of bad ones, and because, in their view, rationally speaking, might makes right. As in Vietnam.

CONCLUSION: LASHED TO THE MAST

In 1970, Robert M. Young correctly reckoned that "the history of brain and behavior research can be seen as the progressive abandonment of faith in a one-to-one correlation between the categories of analysis and the functional organization of the brain on the one hand, and the analogous variables in behavior on the other."[102] The same should be said for the gene and for pharmacology, only too few people seem to know it. "Genetic Determinism," two British geneticists advised *Guardian* readers recently, "is not a concept used by practicing geneticists."[103] Hence, determinist simplicities easily insinuate themselves, with the help of profiting institutions, into public lore.

A social amnesia, to use Russell Jacoby's term, seems to have set in regarding the risks and perils of denying "scientific fallibilism," that is, the sober acknowledgement that "at no actual stage does science yield a final and unchanging result," as Nicholas Rescher writes: "All the experience we can muster indicates that there is no justification for regarding our science as more than an inherently imperfect stage within an ongoing development."[104] "Of course scientific curiosity should be encouraged (though fallacious argument and investigation of silly questions should not)." Noam Chomsky adds, "But it is not an absolute value."[105] The point is not to devalue science but to acknowledge that human interests can be concealed within scientific pronouncements, and to recognize that science occurs in a social sphere and, like it or not, has a social function that can be steered to partisan ends.

The sirens in Theodor Adorno and Max Horkheimer's *Dialectic of Enlightenment* represent overpoweringly sensuous irrationality, but Odysseus's resort to an instrumentally rational tactic, if instead adopted as standard operating procedure, may only end up substituting a different stern songstress to misguide the voyage. Lashed to the mast by his crew, Odysseus listens to the sweet tempting sirens without fear of acting on the dangerous urges they stir. "So instrumental reason can successfully combat myth," as Curtis Bowman notes, "but only at the price of re-establishing a new Myth," that, is a Myth of highly desirable instrumental control.[106] The sirens "also represent a threat to narrative control," that is, a threat to the narrative control of proponents of instrumental reason, which Odysseus temporarily and expeditiously overcomes.[107,]

The trouble is that such a choice is rarely temporary or situation-specific, as in *The Odyssey*. Instrumental reason vies with, and also intersects with, objective reason, or it should. Why then are the sirens in this rendition in distress? They are, after all, not the avatars of objective reason, as Horkheimer and Adorno define it but, lacking the dramatic emotional experience that they offer, one is left with nothing but a desiccated one-dimensional instrumentalist frame.[108] Odysseus has it both ways, and he must in order to be substantively rational in Mannheim's sense or rational in the Frankfurt School sense. The sirens may lure him onto the rocks if they can, but what if he needs to hear, incorporate and, as it were, inoculate himself with their voices to become truly rational so as to navigate more treacherous realms ahead, inner and outer? It seems our shoddy modern versions of Odysseus are determined to plug their ears to the voices of objective reason too.

CHAPTER 9

BACK TO THE FUTURISTS: ON ACCELERATIONISM LEFT AND RIGHT

LANDON FRIM AND HARRISON FLUSS

INTRODUCTION

There are two basic strategies for undermining science: the honest and the subversive. Traditionally, social conservatives have seen scientific knowledge as a threat to their faith-based and hierarchical worldview, substituting a godless nihilism for traditional virtue. American history is full of examples of this "open and honest" conflict. The Scopes Monkey trial of 1925, for example, pitted conservative Christians against the teaching of Darwinian evolution in schools. The bizarre debate between Ken Ham (of the Creation Museum) and Bill Nye, on the biblical flood myth, is a more recent example of the same.[1]

But by the late twentieth century, the Christian Right's strategy began to change. Rather than attack science directly, they attempted to muddy the waters of scientific discovery by expanding the definition of science itself. No longer would science be derided as a worldly tool of the devil. The strategy now was one of subversion. Their rhetoric was aimed at freeing science from secular constraints, i.e., the "secular-humanist" worldview. This change of strategy is well illustrated today by the so-called "Discovery Institute," which advocates the teaching of Intelligent Design, or "ID theory." This institute accuses Darwinian evolutionists of presuming materialism and atheism, while prejudicially excluding the God hypothesis. As one open letter put it, mainstream scientists push "agnosticism upfront but atheism through the backdoor."[2] ID theorists thus position themselves as even more scientific than the mainstream establishment. They feign a radical openness to all hypotheses and pretend not to be wedded to any particular metaphysics. Yet the true purpose of this supposed openness is merely to smuggle in theistic doctrines that they had, in fact, settled on ahead of time.

This strategy of subversion is analogous to another—superficially quite different—tendency relevant today. *Accelerationism* is an aesthetic, cultural, and political trend that seeks to speed up the processes of technology and capitalism in order to radically transform society. Specifically, it has impacted today's political scene, marked as it is by the alliance of Silicon Valley entrepreneurs and right-wing libertarianism—embodied in influential figures like Peter Thiel, the cofounder of PayPal and an early adviser to the Trump administration. Accelerationism, in this right-wing form, gives voice to popular white resentment about economic stagnation and the welfare state but wraps this sentiment up in a shiny, futuristic package.

The term "accelerationism" is often identified with the "neoreactionary" ideas of Nick Land. Through him, accelerationism has influenced the more popular alt-right discourse. Recently, even sections of the political Left have adopted accelerationist ideas, producing works such as the #Accelerate Manifesto, the Xenofeminist Manifesto, and the #AltWoke Manifesto.[3]

Demographically, the accelerationist has little in common with the stereotyped image of the Bible Belt Christian. The former are culturally technophilic, identifying primarily with cybernetics and programming. They emerge from a political economy marked by global world trade, increased automation, off-shoring, and its correlate—the algorithmization of industry (e.g., Uber, Airbnb, and Amazon). Accelerationist buzzwords reproduce the lingo of Silicon Valley: going viral, creative disruption, full automation, and "lines of flight."[4] In their right-wing form, they tend to identify with venture capitalists such as Elon Musk and Peter Thiel, and are sympathetic to right-libertarian politics, which enable all of the above. In their left-wing variants, they fetishize technology for its supposedly Promethean ability to emancipate humanity, rather than a traditional focus on labor.

Whatever their political allegiances, accelerationists mirror the financialization of the economy, where information is cast as the main source of value. Emphasizing free and dynamic creation, speed, and the disruption of established norms, they romanticize science for its liberatory potential. At the same time, accelerationists seek to sever scientific exploration from its proper moorings—that is, an Enlightenment worldview marked by intelligible natural laws.[5]

Enlightenment rationalism is at loggerheads with the accelerationist's insatiable desire to overcome all limits to human and natural existence. Accelerationists therefore adopt the subversive strategy common to Intelligent Design theorists. Rather than opposing science directly, accelerationism poses as its most loyal friend, claiming only to "liberate" science from any constraining metaphysical rules. Repudiated is lawful causality, derived from the *Principle of Sufficient Reason*, and the Cartesian notion of physically extended objects, which exist independently of human perception. In its place is set an anarchic, will-based vision of the world that is radically contingent. Pure potentiality takes the place of comprehensible reality.[6] A dizzying universe marked by unlimited, open possibilities takes the place of an intelligible universe bound by intelligible natural laws.

This accelerationist gambit to "liberate" science is no less harmful than the one affected by sophisticated religious fundamentalists. For the Enlightenment worldview is no straightjacket to science; it is rather its indispensable backbone. In the absence of a lawful universe of mind-independent objects, empirical data loses all its meaning. Such points of data evaporate into nothing more than emotive declarations. What counts as evidence can be endlessly redefined to suit the political program of the moment, blurring the distinction between fact and fancy, and between science and mere faith.

The accelerationist seeks to overturn the Einsteinian/Spinozist vision of the world, wherein "God" is identical to objective Nature, and "divine decrees" are simply the immutable laws of the universe. Einstein's God "does not play dice." But the accelerationist rebellion against such ratio-nalism insists upon something far more romantic, voluntaristic, and in the end, literally super-natural. To Einstein's dictum that "God does not play dice," the accelerationist echoes Niels Bohr's pithy response: "Don't tell God what to do!"[7]

ORIGIN STORY

To understand accelerationist politics, it is useful to trace its nineteenth-century origins and development. Mary Shelley's *Frankenstein* epito-mized this century's synthesis of romanticism and science. Subtitled *The Modern Prometheus*, it depicts a scientist who flouts the mechanical worldview of the Enlightenment. In combining experimentation with the occult arts, Victor Frankenstein escapes both the norms of science and of civilization. In overcoming "ordinary science," he takes the place of God, as the creator of intelligent life, i.e., Frankenstein's monster.[8]

This God-like elevation of human will is likewise the touchstone of Friedrich Nietzsche's philosophy, another key figure in the acceleration-ist's pantheon. Crucially, Nietzsche's infamous declaration that "God is dead," is not so much a theological statement as it is a sociopolitical one—hence its relevance for today's politics. God is dead because "we killed Him" through our popular disbelief. God's death thus signifies the absence of any objective standards or moral authority outside of man himself.[9] We should recognize the sweeping nature of this claim. Nietzsche's death of God not only denies a supernatural arbiter of truth and morality but moreover denies *any* objective standards whatsoever. Only the strong-willed individual is left to create order out of this anar-chic world; they become creators ex nihilo.

The effect of Nietzsche's anti-morality is to condemn the virtues of empathy, compassion, and egalitarianism. Rather than objective moral goods (there are none), these are merely the rationalizations of weak people trying to cope with an inherently hostile, amoral, and chaotic world. Nietzsche subsumes these false virtues under the rubric of "slave morality," and traces their lineage from Rabbinic Judaism to modern democracy and socialism.[10] Only the free spirit, who has overcome slave morality, is fit to affirm life as it truly is, and to rule over those weaker souls that constitute the vast majority. The political program that emerges in Nietzsche is one of aristocratic hierarchy and authoritarianism: The unthinking obedience of the masses is the material fodder for the unguided, free creativity of the very few at the top.[11]

The phrase "accelerate" appears in Nietzsche's late notebooks (the *Nachlass*). Here, Nietzsche anticipates another key feature of accelerationist thought: the idea that progress will be accomplished only through a total crisis and the absolute degradation of humanity. It is the logic of modernity—in all of its decadence—that needs to be exacerbated in order for something new to be born. Nietzsche calls for "accelerating" the leveling tendencies of modern life, to the point where nothing permanent or stable remains.[12] The last men and women of history will be utterly defenseless, obedient, and sheeplike. Out of this degraded ashheap of civilization, new masters arise to dominate the herd from above. Hence, in Nietzsche's *Thus Spake Zarathustra*, the assemblage of mediocre townspeople say to the prophet: "Give us this last human being, oh Zarathustra . . . make us into these last human beings! Then we will make you a gift of the overman!" In other words, the townspeople ask to be subjugated, in order to pave the way for a dominating tyrant.[13]

BIRTH OF A (FUTURIST) NATION

Nietzsche's philosophy of the future subsequently inspired a whole generation of turn-of-the-century and interwar thinkers. These writers reacted to intensified industrialization and the specter of global conflict. While not all explicitly adopted the label "Futurism," they each emphasized a total break from modernity, whether scientifically, aesthetically, or politically. Together they anticipated, and were a bridge to, late twentieth-century accelerationism. In Italy, the standard bearer of futurism was Filippo Tommaso Marinetti.[14] His writings divinized items of technology, from electric lamps to steam engines, factory furnaces, and cars. In many ways, this anticipated the very sort of technology-fetish endemic to technoscience discourse today. Consistently, Marinetti valorized the attributes of velocity, virility, and ecstasy. He even waxed romantic about his car, what he called a "vehement God of steel," in his poem "To my Pegasus." Here, Marinetti personifies his car as rushing "voluptuously . . . into Infinite freedom," and even sexualizes it with the amorous request, "I am at your mercy . . . Take me! . . . Take me!"[15]

Like Nietzsche before him, Marinetti's worldview is scrupulously amoral. In place of any universal, humanistic ethos, there is only the sublime aestheticization of politics. This is particularly evident in the Italian futurist's love for war. For it is in war that a masculine will to dominate is most spectacularly showcased:

> For twenty-seven years we Futurists have rebelled against the branding of war as anti-aesthetic. . . . Accordingly we state: . . . War is beautiful because it establishes man's dominion over the subjugated machinery by means of gas masks, terrifying megaphones, flame throwers, and small tanks. War is beautiful because it initiates the dreamt-of metalization of the human body. War is beautiful because it enriches a flowering meadow with the fiery orchids of machine guns . . . Poets and artists of Futurism! . . . remember these principles of an aesthetics of war so that your struggle for a new literature and a new graphic art . . . may be illumined by them![16]

In Germany, the "reactionary-modernist" Ernst Jünger would extend this line of thought. Whereas Marinetti fetishized modern production and its commodities, Jünger looked to the material basis of these—the modern mass economy with its regimented legions of workers. Retained is the ethos of war and the pervasive amoralism. But with Jünger greater attention is paid to the human subjects that carry out war and turn themselves into the fodder of modern war economies.

This is no humanism: It is not human nature that is celebrated, but rather the subjugation of human beings into something machinic and inhuman. Here we have one of the earliest anticipations of cyborg theory, where the tools of war are not merely adapted to the human form (as with a simple sword), but, on the contrary, the very nerve centers of human beings—their brains and intellects—are molded to seamlessly operate the advanced control panels of weapons systems:

> In order to deploy energies of such proportion, fitting one's sword-arm no longer suffices; for this is a mobilization [*Rüstung*] that requires extension to the deepest marrow, life's finest nerve. Its realization is the task of total mobilization: an act which, as if through a single grasp of the control panel, conveys the extensively branched and densely veined power supply of modern life towards the great current of martial energy.[17]

People are ultimately turned into industrial commodities. Modern capitalism bursts beyond the fetters of its original liberal-bourgeois context. No longer are people the individual bearers of rights, confronting each other in the free marketplace. Jünger's notion of "total mobilization" shifts liberal-capitalism into an authoritarian form: "States transformed themselves into gigantic factories, producing armies on the assembly line that they sent to the battlefield both day and night, where an equally mechanical bloody maw took over the role of consumer."[18] This picture retains the capitalist marketplace while blurring the lines between buyer and seller, what is bought and what is sold. If the nation-state becomes a great war-foundry, then citizens are likewise turned into war-product, to be produced and consumed in the marketplace, now conceived as an international battlefield.

The focus is on technological innovation, divorced from the human frame, and the valorization of speed and efficiency *for its own sake*. As these machinic attributes are foregrounded, the universal human requirements for safety, security, and solidarity disappear into the background. And indeed, the emphasis upon the values of the machine very often comes into direct conflict with basic human decency. When the great spectacle of mass war requires both total mobilization and total destruction, even the notion of innocent civilians must be considered quaint and outdated:

Just as every life already bears the seeds of its own death, so the emergence of the great masses contains within itself a democracy of death. … Giving out the night-flight bombing order, the squadron leader no longer sees a difference between combatants and civilians, and the deadly gas cloud hovers like an elementary power over everything that lives. But the possibility of such menace is based neither on a partial nor general, but rather a *total* mobilization. It extends to the child in the cradle, who is threatened like everyone else—even more so.[19]

What's more, the global scale of conflict that Jünger imagines presumes a rank ordering, not just of individuals or families but of whole peoples and civilizations. Jünger's voluntarism—i.e., his belief in an unbridled free will—accommodates this *weltanschauung* (worldview) of racial division and conflict. For a strong will cannot be bound by universal norms of human equality, but necessarily breaks through these. What's more, there can be no guarantee that such strong wills are evenly distributed throughout the globe.

Belief in such an unbounded will would actually preclude meaningful international agreement.[20] Whereas the Enlightenment rationalist can point to the intellect as the common faculty of all human beings, and a means for coming to common accord, a totally free will is just the sort of thing that defies being "common" or lawful. It is rather inherently unique, anomalous, and practical; every bit as unique as the various cultures of the world.

Jünger saw industrial society as necessarily leading toward a global civil war (with Germany hopefully taking the lead). This would ultimately result in a world-state of worker-soldiers. The problem of nihilism, that Nietzsche had previously diagnosed, would resolve itself on a heroic and planetary scale. Fascist propagandists at the time were indeed influenced by Jünger's ideas of technological world conquest. However, Nazi leaders also thought that Jünger's predictions in *The Worker* were insufficiently racial and völkisch.[21]

Closer to the National Socialist position was the work of Oswald Spengler, a fellow revolutionary conservative in Germany. In his book *Man and Technics* (1931), Spengler speaks of science as a kind of myth-making, or a free and creative pursuit.[22] What's more, the voluntarist nature of scientific praxis means that it will be carried out differently by different peoples. Of all the cultures in the world, Spengler asserts, it is the Germanic "Faustian" culture that most honestly embraces its own self-creation and affirms the dictum that "God is dead." It is therefore amongst the Teutonic peoples that real heroism is most possible; for only they clearly perceive that there is no stable, governing Nature to discover—but only the imperative to assert one's own will:

> True, every scientific theory is a myth of the understanding about Nature's forces, and everyone is dependent, through and through, upon the religion with which it belongs. But in the Faustian, and the Faustian alone, every theory is also from the outset a working hypothesis. A working hypothesis need not be "correct," it is only required to be practical. It aims, not at embracing and unveiling the secrets of the world, but at making them serviceable to definite ends.[23]

This notion of "science as myth" resonates strongly with accelerationists today, particularly in their idea of "hyperstition." Hyperstition is the concept that invented or contrived ideas can have real effects in the world. Doubtlessly this is sometimes true, as in the case of convincing propaganda, virtual crypto-currencies, or even garden-variety advertising. But the idea of hyperstition goes even further than this in its Nietzscheanism. It is the additional claim that all realities are but competing fictions, and that the supremacy of one narrative over another has nothing to do with its relative accuracy, but rather with the agonistic play of competing forces.[24]

On this view, science is necessarily undemocratic.[25] It cannot be based on the inquiry of an objective world by common intellects. On the futurist account, neither exists to begin with. Nor are the fruits of scientific and technological advance brought about with the masses in mind. Rather, scientific endeavor is the heroic act of the strong individual for *their own* practical ends, and *their own* glory:

> In reality the passion of the inventor has nothing whatever to do with its consequences. It is his personal life-motive, his personal joy and sorrow. He wants to enjoy his triumph over difficult problems, and the

wealth and fame that it brings him, for their own sake. Whether his discovery is useful or menacing, creative or distributive, he cares not a jot. Nor indeed is anyone in a position to know this in advance.[26]

This is the very opposite of science as it is normally conceived, namely as the gradual accumulation of facts, and the testing of hypotheses, ultimately for the utilitarian benefit of mankind as a whole. To this picture of "normal science," Spengler shows only disdain. He identifies this as a mediocre sort of technics. This is tied up with a whole history of rationalism and materialism, from the French *philosophe* La Mettrie to the Bolshevik revolutionary Vladimir Lenin.[27]

Rather than technology liberating the masses from toil and drudgery, Spengler sees science as only increasing the necessary demands for labor. For again, there are no natural limits. The machines are insatiable, and they are entirely insensitive to ordinary human suffering or welfare. Not only this, but the titanic scale of industry entails that work will become ever more monotonous. Specialization renders labor incomprehensible to the workers themselves; the narrowness of each job removes its organic connection to the operation as a whole or to any finished product. Not that the true purpose of technology could be understood by mere workers anyway, as this is properly speaking a free and artistic creation of the scientist and the captain of industry.

This whole situation, then, creates a dangerous, reciprocal relationship: The inability of the worker to comprehend her own work, combined with deadening and tedious labor, produces resentment from below. This is matched by an equally strong contempt from above, on the part of the bosses, for a degraded labor force, reduced to unthinking automata. The political corollary to all this is a noxious authoritarianism. C. L. R. James's dictum that "every cook can govern" is necessarily rejected; the lowly worker cannot understand the whole apparatus (whether of industry or the state), let alone have an informed opinion as to its proper operation.[28] Instead, Spengler's technological worldview implies a political despotism that presumes itself unquestionable before a debased, unthinking, and angry mass.

One should not think that this mutual distrust and hatred is merely a "bug" in Spengler's Right-romantic worldview. To the contrary, it is a vital tension in his system, marked as it is by constant struggle and tragedy. Technology atomizes the world. It does this not only by turning skilled craftspeople into assembly-line workers, but at the same time by undermining the traditional institutions of communal life. Much the same has been claimed on the Left, notably in the pages of the *Communist Manifesto*. For in the face of mass production and consumption, the halo is ripped from both family and church. Each individual becomes a producer and consumer in their own right, increasingly drawn into the mass economy and liberated (or alienated) from traditional social structures. As Marx and Engels put it, "All that is solid melts into air, all that is holy is profaned, and man is at last compelled to face with sober senses his real conditions of life, and his relations with his kind."[29]

But, whereas Marxists saw the tensions brought about by technology as a step toward emancipation, the same cannot be said for the fascist-minded futurists. The former see exploitation as giving rise to worker solidarity and consciousness; the latter see the contradictions as simply permanent, destructive, and ever-worsening.

However, it is precisely because of this danger that the right-futurists also saw the place for a world-savior. In the case of Spengler, the final hope for humanity came from Germany, which he saw as the nation least corrupted by materialism, hedonism, and liberalism. An analogous vision for German-led renewal was expressed by the philosopher Martin Heidegger. Never wavering in his appreciation for the "inner truth and greatness" of Nazism, Heidegger claimed that it revealed a new understanding of the "encounter between global technology and modern man."[30]

This all speaks to a pervasive feature of the right-wing German imagination. In wielding new technologies (the radio, the TV, the jet engine, and the V-2 rocket), the National Socialists saw Hitler as forging a new humanity out of the flames of global conflict; for "where danger is, grows the saving power also."[31] This view of global conflict and renewal is at once postmodern and ancient. It posits a heroic savior figure, entirely alien to modern mass politics, and yet using the masses, along with advanced mass production, for his authoritarian goals. After the war, Heidegger distanced himself from Hitler but retained the same messianic hope for the redemption of Europe. As he put it in the infamous *Der Spiegel* interview, "Only a God can save us now."[32]

THE FRENCH CONNECTION

The political influence of futurism was carried into the late twentieth century by a trio of influential French thinkers, Gilles Deleuze, Jean-François Lyotard, and Jean Baudrillard. Nominally on the radical Left, they would use futurist motifs to diagnose the problems of capitalism and modern *anomie*. They have often been associated with the tumultuous events of May 1968, where Parisian workers and students took to the streets and occupied buildings in protest. The fallout from May '68 was a reconfiguration of Left politics and a disillusionment with institutions of the so-called "Old Left," namely the French Communist Party and the Trade Unions. Contemporary theorists have labeled this trio the first of the accelerationists.[33]

Rather than a nationalist authoritarianism, these writers embraced, at various times, libertarian and quasi-anarchist rhetoric. They retained the futurist's eschatology—things are bad; they will only get worse; and it is in the "getting worse" that the last glimmer of hope resides. When it comes to capitalism, there is no point in merely local forms of resistance. Such parochial politics misunderstands the systematic and totalizing nature of the market. The only possibility for emancipation must come from within the logic of the capitalist economy itself. The processes of production, circulation, and inflation should be encouraged maximally. Only by "accelerating" these tendencies can the whole system of capital exhaust itself as it hits up against its own, inherent limits. To pursue an "outside" to capitalism is illusory; capitalism must instead be the engine of its own demise.

As opposed to moralizing against the ruling class from a transcendent position, these writers pinpoint the problem as one of immanent political economy and its pressure points. All of this sounds very materialist. Indeed, these French thinkers sound *almost* Marxist; echoing Marx's dictum that "the barrier to capital is capital itself."[34] This, however, is to ignore their hidden, Nietzschean premise: a strident post-humanism. Affirming the will-to-power involves a simultaneous rejection of any common, human nature uniting all individuals. This human nature, what Marx sometimes calls "species being," would be eschewed as just another illicit "outside" to the market itself. It would be to parochially rely upon some "all-too-human" standard. Gone, therefore, is Marx's vision of a community of associated producers who comprehend the necessities of nature in order to maximize human welfare, leisure time, and material security.[35]

Gilles Deleuze, along with radical psychotherapist Félix Guattari, promotes a metaphysics of desire. Capitalism is (provisionally) good in that it encourages this by constantly revolutionizing production, and thereby demolishing all of the old, static forms of living and producing. The problem for them is that desire becomes "caged," insofar as the present system also throws up authoritarian structures in order to maintain itself. Think of the police, mass incarceration, organized religion, and the bourgeois family. These institutions put artificial breaks on our creativity. So, capitalism "deterritorializes" by atomizing society into individual, anomic producers and consumers; at the same time it periodically "reterritorializes" society by trying to hold everything together through repressive institutions.[36]

The political imperative of the day is to accelerate the process so that we can eventually break through to something new. For them, therefore, schizophrenia is the archetypical modern syndrome—surpassing all authoritarian strictures for an ultimate deterritorialization. While bourgeois society demands the stable ego as the sine qua non of polite society, the schizophrenic instead follows their own uncaged desires.[37]

Not that this was sufficient for all accelerationist thinkers at the time. Jean-François Lyotard would critique Deleuze and Guattari's focus on the libidinal as still retaining some notion of an "outside" to capitalism. For Lyotard, there can be no sense of the libidinal apart from capitalist desire itself.[38] A schizophrenic escape from market oppression will not do; One must rather delight in the "mad destruction" that markets themselves inflict:

> The English unemployed did not have to become workers to survive, they ... enjoyed [*ils ont joui de*] the hysterical, masochistic, whatever exhaustion it was of hanging on in the mines, in the foundries, in the factories, in hell, they enjoyed it, enjoyed the mad destruction of their organic body which was indeed imposed upon them, they enjoyed the decomposition of their personal identity, the identity that the peasant tradition had constructed for them, enjoyed the dissolutions of their families and villages, and enjoyed the new monstrous anonymity of the suburbs and the pubs in morning and evening.[39]

For his part, Jean Baudrillard would go even further, identifying the "terminal velocity" of capitalism, neither in desire nor exploitation but in death itself. The tendencies of markets to suffer inflation, especially, signifies the evacuation of all value from the capitalist system. The post-apocalyptic future of the market is therefore the precapitalist gift economy.[40] This embodies Baudrillard's imagined return to an idyllic past. Premodern values will emerge to replace capitalist exchange value. Foremost of these is the aristocratic ethos of "excess." As opposed to modern economic imperatives to save, and to be efficient, the aristocratic impetus is to expend excess energy in festivals, luxuries, and, generally, the exuberant celebration of life in the struggle against death. This marks an absolute break from Marx's rational critique of political economy, which Baudrillard dismisses as merely "petit bourgeois": "At any rate, Marxism is only the disenchanted horizon of capital—all that precedes or follows it is more radical than it is."[41]

The preceding genealogy of accelerationism is not merely a matter of biography; rather, it outlines certain patterns of thought still present in contemporary politics, art, and science. It should be instructive that these same values have been held by conservative nationalists and post-68 French Radicals alike. For Marinetti and Deleuze, it is an anti-humanist celebration of intensity, the machine, and creative disruption for its own sake. For Jünger and Lyotard it is the identification of alienated workers as the fodder for said machine. Finally, for Spengler and Baudrillard, it is the view that capitalism—the modern economic "machine"—ends with cataclysm, the destruction of liberal norms, and the ironic return to something premodern.

LAND AND THE CCRU (THIS IS ENGLAND)

Accelerationism came into its own during the tech-boom of the 1990s and the early twenty-first century. Responding to the fall of the Berlin Wall, French critical theory, and the 1980s cyber-punk scene, Nick Land, Sadie Plant, and their like-minded graduate students founded the Cybernetic Culture Research Unit, or CCRU for short. The Unit appropriated leftist rhetoric in criticizing bourgeois culture, patriarchy, and heterosexism; at the same time, the CCRU envisioned a futurism that abandoned socialism, or anything that would resemble a Marxist critique of capital. For them, humanism was a disease and man a "drag."[42] Land and Plant's writings emphasized dynamism, flow, the machinic, and an overall vitalism, as well as the celebration of alienation as transformative beyond the merely human.[43]

Sympathetic commentators have called Land's writings an "anti-Logos" program.[44] In other words, his discourses on science and technology reject a rationalist metaphysics. Importantly, Land never offers a direct, metaphysical argument against the *principle of sufficient reason* or any other rationalist tenet. He gives no proofs and makes no deductions. That, in any case, would undermine his own irrationalism and be a performative contradiction. Instead, Land takes a more "pragmatic" and empiricist approach. He invokes current technologies and scientific fads in order to undermine rationalism, without ever engaging with it directly. Quantum mechanics, time travel, virtual currencies, cybernetics, and number theory are all deployed to undermine an intelligible and orderly view of the universe.

Land has a lot in common with some better known philosophers of science, including Paul Feyerabend and even Karl Popper. The latter rejects rationalist metaphysics as necessary for empirical investigation, even taking up the label of "indeterminist."[45] With Feyerabend, the commonalities run even deeper: These include strident statements about philosophy's inability to guide science at all; the denial of any distinction between science and mythic ritual (e.g., astrology and rain dances); a valorization of the pragmatic; as well as a general openness to "incommensurability."[46] This is the idea that two scientific theories may be wholly incomparable since grounded in radically divergent vocabularies, and so no rational choice between them can be made. The social implication is that people holding divergent worldviews may be permanently cut off from any mutual understanding, and also, that the choice between these worldviews is one of will, rather than objective evidence or testing.

Here it is revealing to quote Land at length, especially from his essay *Qabbala 101*:

> Since qabbalism is a practical programme, rather than a doctrine of any kind, its formal errors—mistakes—are mere calculative irregularities, and correcting these is actually a procedural requirement of (rather than an objection to) its continued development. It is the rational dismissal of "the" qabbalistic enterprise that is forced to take a metaphysical stance: ruling out on grounds of supposed principle what is in fact no more than a guiding "empirical" hypothesis.[47]

Note that the mystical worldview Land is invoking here is contrasted, and celebrated, as against the rationalist, "metaphysical" stance. Whereas *qabbala* is a purely practical pursuit, which is self-regulating and self-correcting, rationalism requires something more: namely the belief in indisputable, self-positing, first principles. It therefore claims to be grounded in objective and universal knowledge. The dichotomy Land sets up between theory and practice is highly dramatic. And yet it conforms very well to the typical discourse in contemporary philosophy of science. Namely, the humble, piecemeal, empirical, practical investigator is contrasted to the allegedly dogmatic, overbearing, arrogant, punitive—yet really myopic—rationalist. The rationalist then presumes to make durable rules for the universe without ever bothering to get her hands dirty here on earth.

Again, like Feyerabend, the distinction between science and myth—even the occult—is obliterated. For all such investigations, whatever their original motives, virtuously proceed along strictly pragmatic lines, and so will correct themselves as they go, without the need for any outside, restrictive principles to guide them:

> Epistemologically speaking, qabbalistic programmes have a status strictly equivalent to that of experimental particle physics, or other natural-scientific research programmes. ... There may be no "empirical," procedurally approachable mysteries—or mysterious problems—of the kind qabbalism guides itself towards. If so, it will approach this fact in its own way—empirically, probabilistically, impressionistically, without any logical, transcendental or philosophical meta-discourse ever having been positioned to put it in its place.[48]

Of course, Land gives far too much credit to pragmatic investigation to "self-correct." For in the absence of some grounding, rationalist principles, there can be endless explaining-away of all phenomena that seem to disprove one's core mystical beliefs. But this is nothing particularly new; the whole modern history of evangelical pseudoscience has been an absurd effort of this type. One need only recall the ways that literalist Christians dismiss carbon-dating for the age of the earth, or fossil evidence for dinosaurs preexisting human beings. There is always some supposed flaw or irregularity in the data, the equipment, or the design of the experiment to complain about.[49]

One might imagine that empirical science itself rules out such unconvincing, fringe objections. The mainstream scientists will counter fundamentalist challenges by invoking methodological strictures. They will certify *their* findings, against those of their pious critics, by appeals to the "preponderance of evidence," "statistical significance," or the "repeatability" of their experiments. And they are right to do so. But this is not sufficient. For where is "evidence," "statistical significance," or the import of "repeatability" ever defined? Why are these considered important to begin with? The fundamental problem is this: Scientific methods may be legitimate, but they cannot be self-grounding. That would be viciously circular, and put science on par with any other (internally consistent) doctrine of faith.

On the most fundamental level, the hyper-empiricism of the accelerationist demotes truth itself. For even truth can no longer be considered an overriding, permanent, and grounding value. When metaphysics is set aside, and when one no longer attempts to make sense of an objective world, all that is left is a pragmatic quest for "what works." It may be the case that seeking knowledge simply doesn't "work" as well, or is not as pleasing or convenient as asserting one's own freedom to believe. As Feyerabend states, the choice between incommensurable values will always be something of a free choice.

> And it is of course not true that we have to follow the truth. Human life is guided by many ideas. Truth is one of them. Freedom and mental independence are others. If Truth, as conceived by some ideologists, conflicts with freedom, then we have a choice. We may abandon freedom. But we may also abandon Truth.[50]

We should not pass over this point too quickly. We have here a very revealing statement, and one which matches how the accelerationist mind frames the basic question of human knowledge. Specifically, there is a shift between what we may call a "first order" and "second order" decision. On the first order, there is a decision to be made between ultimate, incommensurable values—truth or freedom. But on the second order, we realize that this choice, itself, is a free and unguided one, entirely a product of the free will itself, rather than evidence or "sufficient reasons." The deck is stacked. A free choice between "truth" and "freedom" will always call the match in favor of the latter, and against the former.

It is then clear how the notion of hyperstition emerges from this basic skepticism about truth and objective standards. In fact, hyperstition is only pragmatism made self-aware. Once we admit that our only criterion is "what works," we are no longer impelled to search for the truth, so much as to invent it. The successful inventions are simply those that have the biggest impact.[51]

While this may strike some readers as consistent with progressive social values, nothing can be further from the truth. The free invention of reality is necessarily antidemocratic. As a product of the creative free will, it implies a battle for domination, rather than a common search for agreement. The questions become: Whose world-narrative comes out on top? Who can shape reality most profoundly?

The implicit authoritarianism underlying the concept of hyperstition is well on display in Nick Land's 2012 work, *Dark Enlightenment*. Here, Land inveighs against liberal, "PC" orthodoxy, what he calls "The Cathedral." Democracy, for Land, is marked by entropy—a consumerist culture that merely reproduces itself incessantly, without ever creating anything genuinely new. Land's excessive use of neologisms help define his political imperative: It is first to transcend the "heat-death" of modern democracy, and then to seek a "bionic horizon" that not only goes beyond the "hubbub" of political deliberation but also redefines human beings as such. The foil to the Cathedral is the "Cracker Factory"—a deliberate play on words, bringing to mind not only anti-PC white culture, but also the physical "cracking up" of the establishment. Land thus combines a postmodern, tech-infused libertarianism with a virulent neoconfederate ideology. Rather than seeking freedom *within* society (as more moderate libertarians might), Land promotes the mantra of "no voice" and "free exit," i.e., secession from modern democracy altogether.[52]

Hyperstition is fundamental to this sort of politics. Land appears to promote a stark realism, counterposing the stale ideologies of liberal democracy to the hard-nosed, empirical recognition of human inequality, including racial disparities in achievement and IQ.[53] But he is not after "hard-truths," not really. In fact, Land's nominalized universe precludes their very existence. His ethos of "cracking up" and "breaking through" is rather about a willful and creative destruction. Land wants to insert new, governing myths in place of the presently hegemonic ones. The bad egalitarianism of democracy, for Land, allows for the lazy masses to leech off of the productivity of creative elites. But when this same egalitarianism breeds resentment, the masses can violently rise up. The solution, in Land's view, is a new ideology of escape—not altogether dissimilar from the notion of a "capital strike" in Ayn Rand's better known novel, *Atlas Shrugged*. But the point is that all of this is myth-making. It is an attempt to reframe reality on different terms than the liberal, the democrat, and especially the Marxist. In particular, it is a ploy to engage people's imaginations, and so to change their minds about the ultimate sources of value. Whereas the Marxist will promote a "labor theory of value," Land recasts workers as merely the passive fodder for novel, techno-industrial creations.[54]

LEFT-ACCELERATIONISM (LEFT BEHIND)

Given the grim history of futurism, might it nonetheless be possible to redeem this tradition for an emancipatory politics? To be sure, attempts to found a "left-futurism" are almost as old as the term itself, and were especially manifest in prerevolutionary Russia.[55] But today, a left-accelerationism is once again en vogue. A myriad of Left blogs, academic symposia, gallery exhibitions, and political manifestos, each drawing on Land's writings, have proliferated over the last decade.[56]

While diverse in their rhetoric and particular aims, these initiatives tend to coalesce around a handful of specific themes. First of all, there is the rejection of a failed folk-politics, i.e., small-scale, non-hierarchical attempts to change whole systems at the local level. This rejection is born out of the failures of the Occupy movement and similar micro-political projects. Second, there is a critique of existing identitarian (racial and gender) politics, which they cast as overly moralistic and given to an ethos of victimization. Instead, left-accelerationists insist upon a return to the universal, and to mass politics. They hold that world systems (such as capitalism) can only be overturned by an equally systematic opposition.

Perhaps the most important example of left-accelerationist politics comes from the work of Nick Srnicek and Alex Williams. In the #Accelerate Manifesto, and in their more recent book *Inventing the Future*, Srnicek and Williams try to synthesize Nick Land's techno-futurism with Marxist ideas of universal emancipation.[57] Instead of Land's praise for authoritarian government and racist overtones, Srnicek and Williams promote a social democratic alternative, with demands for universal basic income, full automation, and technological progress. In contrast to much of postmodern thought, Srnicek and Williams affirm a rhetoric of modernity, rationalism, and the Enlightenment. Against the deep ecologist, they reject a precious worship of nature in favor of an unbridled, Promethean mastery of the environment for human ends.

But in this Prometheanism we can see very clearly the commonalities that left-accelerationism has with the whole bloody history of futurist thought. Specifically, it is the mastery of the Will over Nature that is celebrated and valorized. While left-accelerationists maintain a facade of Enlightenment universalism, they in fact distance themselves from the actual positions of Enlightenment-era thought, especially the determinism and objectivity of Spinoza and the French materialists. Indeed, their supposedly "mass politics" is marked by the same notion of hyperstition as is seen in the neoreactionary works of Nick Land. For them, there is no one intelligible world that can be grasped universally. There is only the agonistic contest of incommensurable worldviews—each considered "universal" from the inside, but nonetheless in constant competition with one another.[58]

This tracks very closely to the left-accelerationist notion of freedom. For them, freedom must be considered a "synthetic enterprise" and never a "natural gift."[59] This is to say that human freedom is not based on some objective and universal view of human nature as such. It is rather something volitional and invented. The rhetoric of Srnicek and Williams around freedom initially sounds conventionally left wing. Freedom is not merely "freedom from" outside restrictions (as the classical libertarian might assert), but is instead the "freedom to" pursue one's desires. But in this, the left-accelerationist tips their hand; for desire, itself, is entirely ungrounded, improvisational, and indeterminate. Who is to say what counts as a proper yearning or inclination? To answer such a question with any certainty at all would require an appeal to something like human nature. But this is precisely what the left-accelerationist rejects. In its place, there is only this aforementioned conflict of worldviews and value systems, that is, incommensurable definitions of desire.[60]

Theirs is at most an existentialist sort of "humanism," or, to use their phraseology, a "humanism that is not defined in advance."[61] The creative will takes the place of any intelligible nature in its open-ended quest for self-definition. Much the same can be said for left-accelerationist takes on racial emancipation and feminism. Drawing on Land's early collaborator, Sadie Plant, they deny knowing what it is to be a woman, and insist instead on an open ended "evolution" of what women's liberation could mean:

> It's always been problematic to talk about the liberation of women because that presupposes that we know what women are. . . . It's not a question of liberation so much as a question of evolution—or engineering. There's a gradual re-engineering of what it can be to be a woman and we don't yet know what it is.[62]

This approach certainly has some distinct benefits. For one, it avoids what Srnicek and Williams term "parochial humanism," i.e., illicitly substituting Eurocentric and masculinist conceptions of the human for a universal essence.[63] Problematically, however, the left-accelerationist lacks any objective values against which to judge their vaunted "evolution" or "engineering" as either progressive or retrograde. In this, they default back into the very sort of parochialism they claim to oppose. For there are only the particular twists and turns of an evolutionary story, and the particular historical moments therein. Never is there a trans-historical standard to evaluate these moments: "The universal, then, is an empty placeholder that hegemonic particulars (specific demands, ideals, and collectives) come to occupy."[64]

For the accelerationist, our "self-engineering" implies a degree of conflict and alienation. We only create ourselves within a particular social nexus. If workers are exploited under capitalism, then this very exploitation is defining of who they are, and who they may choose to become. But here a crucial difference must be noted between the left-accelerationist, on the one hand, and the classical Marxist on the other. For the Marxist, exploitation may well be productive of worker emancipation. If work becomes ever more specialized and centralized, if hours and conditions deteriorate, and if wealth becomes concentrated in ever fewer hands, this may well give rise to class consciousness. Yet this is because such conditions are an assault on a common human nature itself. The need for meaningful work, and social collaboration are part and parcel of what it is to be a human being, and so inhumane conditions may be a spark for class revolt. But the point is that emancipation, on the Marxist view, involves reclaiming the proper life-conditions for human beings, and also developing their innate capacities. This is why emancipation is not only a material development but, at the same time, a moral imperative.

On the left-accelerationist view, however, there can be no human essence, no "species being," considered apart from alienation. Alienation not only defines our circumstances at a particular historical juncture but actually defines the totality of who we are:

> There is no authentic human essence to be realised, no harmonious unity to be returned to, no unalienated humanity obscured by false mediations, no organic wholeness to be achieved. Alienation is a mode of enablement, and humanity is an incomplete vector of transformation. What we are and what we can become are open-ended projects to be constructed in the course of time.[65]

Why one should embark on such a project at all remains an open question. So too does this project's specific agenda. The left-accelerationist claims to promote gargantuan leaps in technology so that the working day can be radically shortened, or even eliminated. Yet one may ask why they would not just as easily take the opposite view, such as the right-futurist Oswald Spengler. As we saw, for him there will be no relief from long working hours; the new, gargantuan machines will simply be manned by ever greater scores of workers, reduced to a state of unimaginable hyper-exploitation. Certainly, the left-accelerationist would abhor such a grim vision, but lacking in humanist principles it is terribly unclear why.

HYPE, HYPE-MAN, HYPERSTITION

Srnicek and Williams have inspired subsequent manifestos that further elucidate the themes of left-accelerationism. These include the Xeno-feminist Manifesto and the #AltWoke Manifesto. The former calls for a rejection of external nature as unjust, in order to make room for the creative will, while the latter embraces a "post-truth" and "post-fact" world to better facilitate Left political designs. The #AltWoke Manifesto takes the cultural ethos of Trump's presidency for granted, in accepting that we are plunged into a world of "alternative facts," which aligns well with the CCRU concept of hyperstition.[66]

There is an even deeper affinity between hyperstition and what we find in something as pedestrian as Trump's *The Art of the Deal*. It may be uncomfortable to left-accelerationists, but hyperstition, the idea that fictions can create realities, comes awfully close to Trump's notion of "truthful hyperbole." Deals can be cut, and projects promoted, through substituting fantasy for the truth:

> I play to people's fantasies. People may not always think big themselves, but they can still get very excited by those who do. . . . People want to believe that something is the biggest and the greatest and the most spectacular. I call it truthful hyperbole.[67]

This notion of willing your way to the top with useful fictions is in fact religious in origin and, at the same time, deeply American. This is the theology of Norman Vincent Peale, that Protestant guru behind *The Power of Positive Thinking*, and Trump family minister.[68] The Trump family's twin mantra of "Be a killer/You are a king," is simply a more vicious version of Peale's "I can do all things through Christ which strengtheneth me."[69] Crucially, Peale's formula is in no way concerned with brotherly love or salvation. Neither is it based upon the actual existence of Christ, but rather the practical benefit of belief itself. Belief—regardless of theological truth—should engender practical success in this world.

A worldview that places practical success over truth will always end up as both unscientific and undemocratic. It will be unscientific because "success" is never self-defining. The pragmatist may appear, superficially, to be a hard-nosed realist who only cares about the facts and results. But what counts as facts or legitimate results will be grounded only in an unaccountable will or decision. Feyerabend's deconstruction of science prevails: Science is indistinguishable from myth or, perhaps, even political propaganda. Democracy is likewise denigrated. For on this worldview collective deliberation about a common world of facts becomes impossible. There can only be a contest of wills, and the domination of the many by the strongest, the most vicious, and, perhaps, the most self-deluded.

In the end, what the accelerationist is up to is nothing terribly new. Despite its futuristic rhetoric, accelerationism merely resurrects an old, religious idea about the absolute freedom of the will, and its ability to create facts ex nihilo. The genuine alternative to this is, likewise, nothing new. Rather, it is rooted in Enlightenment rationalism, which argues for a materialist, secular, and determinist understanding of the universe. On the rationalist view, both science and democracy commence from the very same starting point: an intelligible world. Scientists can engage in repeatable experiments for the very same reason that communities can extend democratic control over their economy. In each case, interpersonal agreement depends on there being an objective, comprehensible world beyond the whims and wills of strong personalities. And moreover, the data that scientists collect should be able to inform communal decision-making, illuminating not only the external world but also helping to further define human needs and flourishing. A democratic society thus needs science to inform its deliberations, but the sort of science adequate to this task must itself be grounded in a rationalist metaphysic.

CHAPTER 10

THE MYTH OF THE EXPERT AS ELITE: POSTMODERN THEORY, RIGHT-WING POPULISM, AND THE ASSAULT ON TRUTH

GREGORY SMULEWICZ-ZUCKER

In April 2016, during a campaign speech, then presidential candidate Donald Trump told an audience in La Crosse, Wisconsin, "You know, I've always wanted to say this—I've never said this before with all the talking we all do—all of these experts, 'Oh we need an expert—' The experts are terrible."[1] In June 2016, British Conservative politician and then justice secretary Michael Gove echoed Trump's sentiment during a Sky News interview in which he supported the leave campaign during the Brexit vote, stating, "people in [Britain] have had enough of experts."[2] These two statements—coming from politicians who have since succeeded in realizing their agendas—speak to the way an attack on experts is an integral part of the global turn to a populist form of democracy. But not all experts are elites, and the association between the very idea of expertise and elitism has become a dangerous myth that has become a valuable rhetorical tool in the hands of populists.

There are two interrelated issues at stake. First, Trump and Gove are not far wrong. The experts, in their technocratic incarnations, have been terrible and people have had enough of them. As the economist Wolfgang Streeck, hardly a partisan of Trump or Gove, has noted, "with the neoliberal revolution and the transition to 'post-democracy' associated with it, a new sort of political deceit was born, the *expert lie*. It began with the Laffer Curve, which was used to prove scientifically that reductions in taxation lead to higher tax receipts."[3] It is everyday people who suffer because of such lies. But this problem has created a second one that has permeated our culture: the justifiable backlash against experts has fostered a deeper distrust of the very idea of truth. After all, experts claim privileged access to the truth. The populist forces that imperil democracy have employed the critique of experts as a way to assault truth. This has devastated the status of science in our society. The neoliberal economic expert gets painted with the same brush as the climatologist. For example, Daniel Dennett, discussing debates over intelligent design (ID), has observed:

> Biologists have been quick to respond, issuing incisive rebuttals to the various claims about the scientific integrity of ID, but these denunciations create the impression that an elitist scientific establishment is smothering an underdog, and a virtuous and plausible one at that—a theory quite literally "on the side of the angels."[4]

The content of the scientist's claims becomes irrelevant because what matters is their status as purported elites. The fact that most biologists are experts of a very different stripe than the technocrats advocating austerity programs, and, consequently, hardly elites in any meaningful sense, is beyond the point. All expertise becomes synonymous with elitism and, thus, truth—that with which the expert is supposed to concern herself— is seen as a tool of the elite.

In the populist rhetoric of our time, the assault on experts as elites has become cloaked in the language of democracy. This is an ill that affects left-wing discourse no less than right-wing discourse. In the sections that follow, I shall offer synoptic accounts of how the assault on expertise feeds into an assault on truth and advocacy for a populist conception of democracy on both the Left and the Right. However, it is crucial to keep in mind that on the Left the assault on truth has largely come out of the realm of a convoluted postmodern theoretical tradition that has done more to stultify Left politics. In contrast, the Right has enjoyed alarming success in building a political movement that has relied heavily on the assault on truth. Though my discussion of right-wing activism is largely focused on developments in the United States, it would be a mistake to ignore the ways this attack on truth has fueled the global shift to the populist Right.[5] I conclude by proposing that both science and democracy retain values with practical import that can respond to the extremes of technocracy and populism.

THE POSTMODERN ASSAULT ON TRUTH AND EXPERTISE IN THEORY

Ironically, laying blame on postmodernism for just about everything has become a favorite pastime of both the defenders of science as well as the resurgent right-wing critics of expertise. The defenders of science get the critique of the epistemological assumptions broadly correct, but they also miss the political implications of the ideas. In contrast, for its right-wing critics postmodernism has less to do with the substance of the ideas. Rather, postmodernism is a euphemism for all that is wrong with academic elites who employ nonsensical jargon. To the extent that they comprehend the ideas, they object to its relativistic position on values. Yet, what many of them miss is that postmodern theory is perhaps the most sophisticated defense of their brand of populist democracy. Postmodern theory is, perhaps, the theory the right-wing political activists need.[6]

While constituted by a byzantine set of theorists and theories, three thinkers appear to me to be especially relevant to explicating the anti-science and populist undercurrents of postmodernist thought: Michel Foucault, Paul Feyerabend, and Ernesto Laclau.

Foucault, perhaps the best known and most widely cited of the three, is one of those rare thinkers one is likely to encounter in nearly every discipline. The central theoretical claim that runs throughout his work, that knowledge claims are imbued with power relations, speaks directly to the assault on expertise and the notion that experts are elites. As Foucault puts it,

The important thing here, I believe, is that truth isn't outside power, or lacking in power: contrary to a myth whose history and functions would repay further study, truth isn't the reward of free spirits, the child of protracted solitude, nor the privilege of those who have succeeded in liberating themselves. Truth is a thing of this world: it is produced only by virtue of multiple forms of constraint. And it induces regular effects of power. Each society has its regime of truth, its "general politics" of truth: that is, the types of discourse which it accepts and makes function as true; the mechanisms and instances which enable one to distinguish true and false statements, the means by which each is sanctioned; the techniques and procedures accorded value in the acquisition of truth; the status of those who are charged with saying what counts as true.[7]

Gone is the notion of any objective conception of truth. Yet what is significant about Foucault's argument and makes it so thoroughly damning for the notion of truth is the notion that truth is permeated by power. It's not simply that elites or technocrats make use of their knowledge to promote their agendas, which would constitute a critique of technocracy. Rather, it's that each and every knowledge claim entails an ordering and categorizing of society that is at one and the same time a disciplining of society. It is for this reason that a large part of Foucault's work shows a special interest in the way historically specific manifestations of knowledge categorized mental illness and institutionalized people as a result (and, as an aside, I might note that I am not entirely unsympathetic to the significance of Foucault's findings; it his deeper epistemological claims to which I object). However, Foucault confuses the historical manifestations of scientific practices with the very nature of science itself.

Hence, there is no objective truth in Foucault's theoretical schema, only regimes of truth. Of course, the scientist is a natural target for Foucault because of her reliance on objectivity and sees an intimate entwinement between the birth of the modern expert and a politics that entails the disciplining of subjectivity. He writes,

The "specific" intellectual derives from . . . the savant or expert. . . . No doubt it's with Darwin or, rather, the post-Darwinian evolutionists that this figure begins to appear clearly. The stormy relationship between evolutionism and the socialists, as well as the highly ambiguous effects of evolutionism (on sociology, criminology, psychiatry, and eugenics, for example), marks the important moment when the savant begins to intervene in contemporary political struggles in the name of a "local" scientific truth—however the important the latter may be. Historically, Darwin represents this point of inflection in the history of the Western intellectual.[8]

So within Foucault's thoroughgoing assault on truth there is an important place for an attack on expertise, especially the expert in the natural sciences, as tied to the way society is ordered. In a sense, then, there are two levels of Foucault's attack on objectivity. The first is the obvious issue of the expert whose claims have an impact on the ordering of society. Of course, there is much historical and contemporary evidence to back this up. Foucault is not wrong about this. The second, more radical, claim, that truth is imbued with power, goes a step further. It is this claim that is consonant with our world of "alternative facts" and the idea of the "post-truth society." It is this claim that is consonant with the fragmentation of society into private spheres where, since all truths are expressions of power, each becomes equally valid.

In one sense, Foucault's argument can be utilized as a critique of the powerful and their claims. This is what the Left has most commonly picked up on in his thought. Yet, at its deeper level, Foucault's argument leaves one without any objective basis to challenge the powerful. Your truth is just as imbued with the attempt to order society as their truth. So it becomes a competition of truths with no appeal to some objective state of things. It is in this respect that current encouragement to get individuals to "speak their truth" is so irksome. It is a kind of trickling down of Foucauldian ideas into the popular parlance. It contains within in it the notion that there are, indeed, multiple truths, and no one's account of the truth should be conferred special distinction. What is important about the theory is the kind of commitments people are automatically making when they invoke it—commitments to a highly relativistic conception of truth that denies its very importance as anything other than wholly subjective. At one and the same time, truth's relation to power becomes a tool for the critique of power as well as a defense of my private truth. Further, my private truth becomes a source of my empowerment when I oppose your truth. The implications for democracy become worrisome because it becomes a way to doubt objectivity as well as a way to valorize the truth embraced by factions, with no prospects held out for reconciliation or the correction of incorrect views. The role of the expert as an authority is totally denigrated.

Such implications of Foucault's thought are carried forward in the philosopher of science Feyerabend's more direct discussion of the relation between science and democracy. Building off of his more theoretical work, Feyerabend argues for a purported privileging of democracy over truth. In an argument that ought to appeal as strongly to the libertarian as the anarchist (which Feyerabend saw himself as), he writes:

> In a democracy an individual has the right to read, write, to make propaganda for whatever strikes his fancy. If he falls ill, he has the right to be treated in accordance with his wishes, by faithhealers, if he believes in the art of faithhealing, by scientific doctors, if he has greater confidence in science. And he has not only the right to accept, live in accordance with, and spread ideas *as an individual*, he can *form associations* which support his point of view provided he can finance them, or find people willing to give him financial support. This right is given to the citizen for two reasons: first, because everyone must be able to pursue what he *thinks* is truth, or the correct procedure; and, secondly, because the only way of arriving at a useful judgment of what is supposed to be the truth, or the correct procedure is to become acquainted with widest possible range of alternatives.[9]

Such claims make sense as the only conception of the relation between science and politics if one follows the political implications of Foucault's theory. Indeed, Feyerabend works within the framework of the same leveling of "truths" as Foucault's theory implies. And Feyerabend's extreme individualism leads him to pit the individual against truth. For this reason, and building off of his claim that the real advances of science came from outsiders who resisted expert opinion, Feyerabend advocates the democratic oversight of decisions in the sciences by laypeople: "If the taxpayers of California want their state universities to teach Voodoo, folk medicine, astrology, rain dance ceremonies, then this is what the universities will have to teach. Expert opinion will of course be taken into consideration, but experts will not have the last word. The last word is the decision of democratically constituted committees, and in these committees laymen have the upper hand."[10] Such an argument—based on a conception of extreme individualism's logical conclusions—is far more persuasive a grounds for teaching intelligent design or refusing to vaccinate your children than any fostered by either movement's supporters. And it, much like Foucault's ideas, is built upon an argument against the objectivity of truth. Truths are decided upon, not discovered.

Neither Foucault nor Feyerabend have a thoroughly worked out theory of politics. Rather, we only get a dim sense of what kind of politics either would endorse based on their theories of knowledge. For both, it would involve just about anything that would challenge prevailing norms and social orders. Yet what is important to emphasize here is that their opposition is not focused on regime types or political systems. Rather, it is against systems of knowledge. That, for them, is where the oppression lies. Not in authoritarian governments or capitalism run amok. So this raises the question as to what kind of politics postmodernism can ultimately endorse.

It seems to me that the clearest formulation of a postmodern politics comes out of the work of Ernesto Laclau. Along with his frequent collaborator Chantal Mouffe, Laclau sought a conception of politics that could unify the idea of difference that lies at the heart of postmodern politics.[11] He found it in a theory of populism. In a world defined by extreme differences, what can unite people amidst their differences so that they can engage in the kind of coordinated action necessary to politics? For Laclau it is the idea of the people. As Laclau explains, "a certain identity is picked up from the whole field of differences, and made to embody this totalizing function."[12] For a social world in which difference is the prevailing characteristic, something needs to stand for the unity of the group, i.e., embody the totality. Yet, for Laclau, it is a unity that does not dissolve differences because it is more abstract idea than concrete manifestation of an alliance of political interests. "The 'people,'" Laclau explains, "is something less than the totality of the members of the community; it is a partial component which nevertheless aspires to be conceived as the only legitimate totality."[13]

For Laclau, populism is the true essence of politics. As Laclau argues, "Since the construction of the 'people' is the political act *par excellence*—as opposed to pure administration within a stable institutional framework—the *sine qua non* requirements of the political are the constitution of antagonistic frontiers within the social and the appeal to new subjects of social change—which involves, as we know, the production of empty signifiers in order to unify a multiplicity of heterogeneous demands in equivalential chains."[14] The idea of the people is a kind of necessary social construction to make up for the fact that social life is defined by difference. It becomes the most encompassing political category because it reduces politics to its essence, antagonism, us versus them. This lends itself to a conception of politics as the people versus the elite. However, elites need not be identified by any particular characteristic. The "they" are whoever works within the framework of institutionalized political practices.

Much leftist thought that stands in opposition to postmodernist thought bases itself on a clear conception of who the elites are, what their interests are. They are clearly identifiable based on who actually wields material, concrete power. For Laclau and his postmodernist compatriots, there is no analytically precise definition of what makes someone an elite. It is simply the "they" who stand in opposition to the "people." What exactly it is that someone stands for and the reasons why they stand for it cease to matter. Politics is reduced to its barest us versus them idea. And the "them" is anything that seeks to homogenize society. The "people" become an abstract symbol—an "abstract signifier" as Laclau puts it—that represents a thoroughly differentiated society. It is a symbol of the totality without actually being a concrete unity because it is the most encompassing category. It is not encompassing because it is pluralistic. Rather, it is encompassing because it defines a break between two great warring camps, the elites and the people.

Using Foucault, Feyerabend, and Laclau as my reference points, I am suggesting that there is a link in postmodern thought between the rejection of truth and supposed celebration of difference and a certain conception of populism. At issue is the way the loss of any conception of the value of reason in binding society together paves the way for a conception of politics in which our social bonds are based on abstracted notions. Moreover, what is evident in all three thinkers is a disdain for an elite, but an elite that has to be abstractly defined because it is the truth-regime that is the locus of power rather than particular agents or groups. Indeed, one of the things that make these arguments so interesting for the purposes of the discussion in this volume is that the only way these thinkers can arrive at such claims is by divorcing their analyses from anything grounded in objective structures.

What has escaped the purview of critics of postmodernism coming out of the natural sciences is that they tend to miss the fundamental nature of the claim about knowledge that informs postmodernist thought. They are not merely advocating social constructivism, relativism, or, for that matter, celebrating irrationalism. Nor are they simply making a claim for the expansion of democracy in the name of difference. This is to subsume their thought in categories that are perhaps too familiar. On the contrary, they are fundamentally hostile to the notion of knowledge because it entails oppression, the splitting apart of an inherently differentiated world into clean categories. Indeed, many of postmodernism's critics, on both the Right and the Left, mistakenly argue that the postmodern conception of power is grounded in the idea that there is an interest behind knowledge production. In fact, the point, as particularly evident in Foucault, is that knowledge itself is a form of power. Postmodernism divorces itself from any concrete conception of interest. Knowledge itself is disciplinary. There is no place for interested agents in the theory. This is the critical move that makes it impossible to try to reconcile the kind of thinking that those with sympathies leaning toward the natural sciences embrace with the kind of thought espoused by postmodernism. It is also the kind of thinking that leads Laclau to embrace that most nebulous category, the "people," which stands in opposition to an equally nebulously defined elite.

While such jargon-laden talk has been influential among theorists on the Left, it is also true that it has had little actual political impact. Such texts generally do not reach audiences far beyond graduate programs in the humanities and social sciences or select reading publics in urban centers. Postmodern theory is hardly the kind of thing that gets people marching in the streets. If it were, people would not even know what it is they are marching against. There is every reason to worry about the influence of these texts, but to say that they extend beyond high-academic circles would be to exaggerate their political impact. If their impact has been anything, it has largely been the paralysis of Left political movements. Postmodern theory has done more to push the Left into an intellectual morass than to inspire political activism with any clear agenda. Therein lies the cause for concern. Certainly, this has a deleterious effect on educational systems and has plunged debate into the realm of the nonsensical. It has, to adopt a distinction articulated by the political theorist Jon Elster, produced more waste than harm.[15] Perversely, the political manifestation of the high theory espoused by the postmodernists has asserted itself most clearly on the Right.

THE RIGHT-WING ASSAULT ON TRUTH AND EXPERTISE IN PRACTICE

Going at least as far back as the Counter-Enlightenment, a critique of intellectuals has long been a hallmark of political conservatism. Conservatives have often defined themselves in terms of their skepticism of intellectuals who they think have sought to undermine values that stabilize society.[16] Their talk hinges on a defense of Western civilization against philistinism. Thus, in the mid-1960s, Russell Kirk could write, "At this hour when Communists and other totalists are busy ripping to shreds the 'wardrobe of a moral imagination,' certain people of a different cast of mind have turned tailors, doing their best to stitch together once more the fragments of that serviceable old suit we variously call 'Christian civilization' or 'Western civilization' or 'the North Atlantic community' or 'the free world.' Not by force of arms are civilizations held together, but by the subtle threads of moral and intellectual principle."[17] It is the kind of worldview that made traditional conservatives, like William F. Buckley, rush to attack scientists and support creationism as a way to defend Christian civilization.[18] But the guarding of tradition and the kind of conservatism that people like Kirk and Buckley built, complemented its assaults on intellectuals with its own intellectual elitism. They tried to stave off the Far Right in their mid-twentieth-century efforts to reconstruct the conservative tradition.[19]

The contemporary political Right of people like Steve Bannon has revolted against the elitism of that generation of conservatives in favor of a right-wing populism. Still, the trope of the critique of the intellectual in the name of defending established, unchanging values remains a part of the rhetoric of the current Right. In this respect, Kirk and Buckley's worldviews undoubtedly persevere. Nevertheless, the internet has assuredly changed everything. Just a little under a decade ago, the internet was being hailed for its democratization of knowledge. Leftists readily rejoiced at the work of hackers seeking to foment chaos for the "lulz," thinking this exemplified democracy in action.[20] On the contrary, the internet has emerged as an excellent tool for feeding right-wing paranoia, developing its own world of truths, and serving as device for uniting people with fringe beliefs or recruiting people to Far-Right causes.[21]

The internet has cultivated a world where wildly popular right-wing provocateurs like Anne Coulter, Milo Yiannopoulos, and the psychologist Jordan Peterson, climate-change deniers, or the so-called "citizen journalists" of Breitbart News reign freely. Some engage in the dissemination of patently false information, while others, particularly Coulter and Yiannopoulos, simply revel in the pleasures of sophomorically thumbing their nose at authority by tossing around insults. Yet undergirding all of these activities is a worldview that bears a striking resemblance to that of postmodern theory. Whether promulgating misinformation or tweeting racist, sexist, or homophobic comments, all of these people see themselves as engaged in a battle against regimes of truth perpetuated by shadowy intellectual cabals trying to discipline society.

Despite the fact that postmodern theory as taught in some university humanities departments is often their target, they are themselves, in fact, postmodernists without the sophisticated epistemological theory. They are vulgar postmodernists. I say vulgar because they claim to identify an actually existing elite, when their definition of the elite is as nebulous as that of postmodernism. Yet this fiction has helped inspire right-wing populism's myriad conspiracy theories, and it is under the influence of such conspiracies that the modern Right is mobilized. Because examples of this are too many to catalog in the space of this chapter, allow me to focus on two phenomena that illustrate the way the internet has facilitated the promotion of a right-wing conception of truth regimes that has also fueled political mobilization.

The frenzy over a purported malevolent reshaping of the meaning of truth informs the notion of being "red-pilled," which has become part of the cryptic parlance of the alt-right. A reference to the film *The Matrix*, in which the protagonist takes a red pill and becomes aware that the world he lives in is an illusion constructed by devious computers, being "red-pilled" refers to being awakened to the truth and converted to a right-wing world of conspiracy theories.[22] Young men are often "red-pilled" on the internet after searching for relationship advice. On websites like Reddit, they are exposed to a host of conspiracy theories that attribute their relationship failures to a feminist agenda to emasculate men. This, in turn, often quickly leads to venomous misogynistic postings on community message boards and the promulgation of conspiracy theories about Jews, African Americans, global elites, and the standard cast of shadowy figures manipulating the world order.

In an environment where disillusioned young men scour the internet for answers, it is no accident that the Canadian psychologist Jordan Peterson has found an enthusiastic audience for his many YouTube postings among the alt-right as he inveighs against the supposed machinations of women's studies programs, postmodern theorists, and so-called cultural Marxists (by which he actually means the Frankfurt School of critical theory, whose members, Peterson's acolytes are often quick to observe, were mostly of Jewish descent).[23] All the more ironic then that such visions of a constructed truth regime, as allegedly promoted by women's studies departments and cultural Marxists, sounds so much like Foucault's regimes of knowledge used to oppress and reorder society. The truth that red pillers come to recognize is a quasi-spiritual epiphany that everything they thought they knew is false. This "truth" is totally barren with little to support it other than the bare assertion that there is an elite trying to reshape the way you think as a way of disciplining society.

Like red-pilling, the rise of so-called "citizen journalism" is a phenomenon linked to the rise of the Far Right and has been well-cultivated by the internet.[24] Websites like Breitbart News and Alex Jones's InfoWars decry the pernicious influence of the Mainstream Media (MSM), encouraging people without any journalistic credentials (and part of having solid credentials as a journalist entails an education in professional ethics and methods) to conduct their own investigations.[25] Once again, one enters a world of conspiracy theories where unchecked sources are used to substantiate claims about destroyed election ballots and Hillary Clinton's membership in a child molestation ring based in the backroom of a pizza parlor. The rise in popularity of such right-wing journalists/political agitators has only been enhanced by the power of social media, where outrightly false stories are more likely to circulate than stories by established news outlets.[26] Add to this the promotion of such stories by Fox News and a president who dismisses any story that is critical of him as "fake news" and we have arrived at a situation where people see any story coming from authoritative news sources as propaganda promoted by a shadowy elite.

The rise in popularity of notions like red-pilling and citizen journalism represent a major shift in right-wing assaults on the idea of truth. Just a few years ago, the assault on objective knowledge from the Right appeared more focused on the natural sciences, especially the teaching of evolution or the fact of climate change. However ridiculous and unqualified opponents of the teaching of evolution and so-called climate skeptics are, they tried (and continue to try) to promote their ideas by proposing competing theories.[27] Such were the tactics behind absurd claims about intelligent design or historical cycles of climate change. They were trying to beat the scientists at their own game, albeit through what, more often than not, seemed like a deeply distorted understanding of how scientific inquiry is practiced. Today, the situation is much worse. The very concept of truth is treated as an ideological construct. Truth is increasingly treated as the enemy of the people. It is seen as meant to discipline the common person. It is taken as an affront to established beliefs and as part of a campaign to humiliate people's intelligence so that they will blindly accept what they are told. This is the point where the equation between power and knowledge and populism meet. If arguments based on fact are associated with an elite cadre of experts, then it is only that nebulous force, the people, which can stand up in opposition to it.

Some may take exception to my focus on the Right in the assault on knowledge. Of course, many on the Left have long been susceptible to conspiracy theories. There is also plethora of seemingly apolitical forces promoting falsehoods. The claims of anti-vaccination activists and self-help gurus are no less culpable in cultivating a culture where one person's opinion is as valid as any other. Undeniably, the crisis of truth and the increasingly fragile position of the expert is part of a broader social crisis. Yet the Far Right has distinctively positioned itself to turn the assault on truth into a political force. Gwyneth Paltrow's absurd and dangerous claims about women's health[28] do make her as harmful as an Alex Jones. Deepak Chopra[29] and Jenny McCarthy[30] make claims about science that are as uninformed as any climate-change denier. The popular left-wing theorist Slavoj Žižek's pseudo-theories[31] are no more open to actual objective inquiry than the creationist's theory of irreducible complexity. The difference is that the Right has carried this anti-truth and anti-science project in a direction where it has become a genuine threat to democracy.[32]

It is the Right that has built a political movement around falsehoods. Trump's "Make America Great Again" slogan has become a rallying cry for the people to rise up against the elites who purportedly peddle in "fake news." Indeed, part of Trump's continued fixation on the number of attendees at his inauguration betrays the authoritarian populist's need to show that his will is identifiable with the authentic will of the people. In the populist distortion of democracy, the will of this nebulously defined "people" dominates all else. This is the actual realization of what it means when one, as Laclau does, approvingly talks about the people as "abstract signifier." Were he alive, Laclau would certainly have objected to the brand of populism now running rampant across the globe. It is often pregnant with xenophobia, racism, and misogyny in ways that Laclau would undoubtedly oppose. Yet, that is precisely the problem with "abstract signifiers." They can be made to signify anything. Consider, for example, the white nationalists marching in defense of Confederate monuments in Charlottesville. Regardless of the fact that these monuments were erected decades after the Civil War in the context of the reasserted racism of the Jim Crow era, it is a more benign Southern cultural heritage that the apologists for such monuments claim that they signify.[33] It's the peoples' word against that of the historians.

So what explains the rise of the assault on truth by the Right? Certainly, the internet has proven an exceptional mechanism for the dissemination of falsehoods and the Right has become more expert at using it to mobilize and indoctrinate people. There are also clear vested interests in opposing the findings of natural scientists, especially with respect to climate change, on the part supporters of certain economic interests and religious beliefs. Still, this does not explain the appeal. Ultimately, the views disseminated on the internet or funded by wealthy elites have to resonate with the population at large.

The German sociologist Max Weber famously argued that modernity, with its increasing emphasis on rationality, entails the "disenchantment" of the world. Science and its focus on objectively valid claims is one of the major contributing factors to this disenchantment. It dismisses belief and dogma in favor of rationally grounded ideas. But this is an old story that goes to the advent of the Scientific Revolution and the Enlightenment. It was an emphasis on the superiority of reason versus belief that animated opposition to the Enlightenment by established authority, whether of the church or the crown. It was also precisely this authority of reason that postmodern theory, in its various stripes, sought to dethrone.

In a certain respect, there is nothing new about the assault on truth that is occurring today. Take for example the response of Derwin, a Louisianan interviewed by Arlie Russell Hochschild, to the industrial pollution of the Bayou d'Inde: "In the Garden of Eden, 'there wasn't anything hurting your environment. We'll probably never see the bayou like God made it in the beginning until He fixes it himself. And that will happen pretty shortly, so it don't matter how much man destroys."[34] The recent director of the Environmental Protection Agency, Scott Pruitt, shares these sentiments.[35] Such viewpoints are echoes of William Jennings Bryan's response to Clarence Darrow while under examination during the infamous Scopes Trial. Bryan told the court, "The purpose [of Darrow's examination] is to cast ridicule on everybody who believes in the Bible, and I am perfectly willing that the world shall know that these gentlemen have no other purpose than ridiculing every Christian who believes in the Bible."[36] In terms of belief, there is little separating Derwin and Pruitt from Bryan.

What has changed is the scope of the assault on truth. In our thoroughly technologized world, the internet has proven an excellent tool for expressing opposition to truth. Opposition to the idea of truth never went away, but it has now gained a new platform. One of the most disturbing and fascinating things about the online world of the Far Right is its use of symbols. With something as ridiculous as a cartoon frog named Pepe, the Far Right has revitalized a world where symbols become imbued with deep meanings. Even the term "red pilling," with its reference to a science-fiction movie, is a striking example of how a vocabulary derived from the world of fantasy is utilized to describe reality. However inane these examples are, they are indicators of a kind of "re-enchantment" of the world. What matters is the fact that this jargon and these symbols are so deeply invested in by the people who use them to make sense of the world. Pepe the frog is no less an abstract signifier for the people opposing a phantom-like elite than Laclau's "people." It is a world in which evidence that does not appeal to people can be discounted as alternative facts promoted by the mainstream media.

Postmodern theory and the contemporary politics of the Right have an elective affinity. With the condemnation of truth as any kind of objective referent to how the world is, the only kind of democratic politics that can emerge is one in which everyone is governed by belief. Today, out of dismissal of truth, has arisen a world where belief governs political decisions. The Right, in its long emphasis on traditional belief, finds a natural base among people who see the authority of reason as a threat to belief. While I very much doubt that Kirk or Buckley would endorse the politics of the contemporary Right when they set about reinventing conservatism, contemporary conservatives of their ilk, like Nicholas Kristol, John Podhoretz, and the late Charles Krauthammer, readily cringe and have come out against this new wave of the Right. But the invective of a Steve Bannon against establishment Republicans may be a case of the revolution coming to eat its own. This is a base that conservatives have long cultivated and their turn toward displaying themselves as the protectors of Kirk's "Christian civilization" gave them a base that helped them win elections. As Ronald Reagan famously told evangelicals, "I know you can't endorse me. But I endorse you."[37] Indeed, invocations of the people were already evident in Richard Nixon's talk of the "silent majority," which was later appropriated by the Trump campaign.

TOWARD A RENEWAL OF THE SCIENTIFIC ETHOS AND THE DEMOCRATIZATION OF EXPERTISE

I have raised two concerns in this chapter. The first is that there is a justifiable suspicion of the experts. The second is that this justifiable distrust has paved the way for a populist assault on truth. The solution to both problems lies in renewing the promise of science and democracy.

"Science," the physicist Richard Feynman suggested, "is the belief in the ignorance of experts."[38] Feynman's definition sides with that great tradition of modern science as a form of resistance to established belief. Ostensibly, this is a definition that could be vacuously invoked by Feyerabend or a climate denier in defense of what they are doing. After all, they believe the experts are ignorant. However, Feynman's definition is not a naive invocation of the value of doubt. An apt warning precedes his definition of science:

Science . . . teaches the value of rational thought, as well as the importance of freedom of thought; the positive results that come from
doubting that the lessons are all true. You must here distinguish . . . the
science from the forms or procedures that are sometimes used in developing science. It is easy to say, "We write, experiment, and observe, and
do this or that." You can copy that form exactly. But great religions are
dissipated by following form without remembering the direct content
of the teaching of the great leaders. In the same it is possible to follow
form and call it science but it is pseudoscience.[39]

We ought to question the experts, but science has a particular reliance
on the use of rational thought. Science discriminates against ignorance,
whether of the expert or the demagogue.

The rhetoric of our sociopolitical climate applies Feynman's definition unevenly and discriminately. In believing in the ignorance of experts,
people have failed to believe in the ignorance of the demagogues and
pseudoscientists. Science is an inherently democratic enterprise in that
the reasoning and procedures that undergird it are available to all. It is
in its self-conscious exposure to the doubts of others, by making its reasoning and procedure transparent, that a scientific claim can be subject to
correction. The same cannot be said of the claims of those who deny truth
and, consequently, do not have any reasons or procedure that substantiate
their claims. It is when no reasons are given that we should suspect that
ignorance, in its most pernicious form, is governing.

Recovering that scientific ethos that seeks to expose the ignorance
of authority is crucial as a bulwark against the technocrats, dogmatists,
and populists. However, experts still play a vital role in our society and
it is crucial that trust in experts be restored because of all the good they
can do.[40] This trust, however, must be warranted. Here, a kind of renewal
of the democratic ethos is necessary. Democratizing expertise does not
mean setting up Feyerabend's non-expert citizen review panels. Rather,
it means having experts who are actually responsive to the issues they are
supposed to address. The public perception that the experts have become
a self-contained class is not incorrect. In his critique of contemporary liberalism, the social critic Thomas Frank has pointed to the history of the

Democratic Party as a party that has severed its ties with working people and increasingly become the party of the professionalized and managerial elites. As Frank observes, "Liberalism itself has changed to accommodate its new constituents' technocratic views. Today, liberalism is the philosophy not of the sons of toil but of the 'knowledge economy' and, specifically, of the knowledge economy's winners: the Silicon Valley chieftains, the big university systems, and the Wall Street titans who gave so much to Barack Obama's 2008 campaign."[41] Indeed, when Trump and Gove raged against the experts, they had such people in mind.

In opposition to the model of turning to the experts of the professionalized class for guidance, Frank looks to the example of Franklin Roosevelt's administration. "Unlike the Obama administration's roster of well-graduated mugwumps," Frank writes, "the talented people surrounding Franklin Roosevelt stood very definitely outside the era's main academic currents."[42] This paved the way for a group of heterodox innovators who sought new solutions to the country's woes. The ignorance of experts, like Andrew Mellon, the wealthy banker turned Treasury Secretary who counseled Herbert Hoover to do nothing in response to the Great Depression, was recognized. This rejection of an established elite based on their failure to provide solutions to real problems is an example of both the democratization of expertise and the scientific ethos in practice.

One implication of everything I have suggested is that one of the great affinities between science and democracy is that both practices rely on the challenging of illegitimate authority. The wrong claims and bad administrators get weeded out. When science and democracy fail to perform these functions, there is every reason to retreat into the worlds of irrationalism, myth, and demagoguery. The scientists have proven more skilled at fending off bad ideas than the politicians. But it is possible to make a connection between how loss of public confidence in the experts of one sphere can turn the public against experts in another.

As the historian Richard Hofstadter argued over fifty years ago, a certain anti-intellectual strain has been a persistent feature of American life.[43] There is little reason to think that it will go away. Nevertheless, it is possible to prevent it from embedding itself in the broader norms of our society and the values of our political institutions. Scientific and democratic values can play an urgently needed role in combating the populist tendencies of our age, as well as providing us with the experts we need.

PART IV

THE REVENGE OF ANTI-SCIENCE

CHAPTER 11

PLATO'S REVENGE: AN UNDEMOCRATIC REPORT FROM AN OVERHEATED PLANET

PHILIP KITCHER

I

By midsummer 2099, as yet more record temperatures were reported from an ever wider collection of places, the intergovernmental office of environmental studies finally announced the answer to a long-debated question.[1] Many figures for the rise in the mean temperature of the earth during the twenty-first century had been predicted, but it turned out that the right value was 5.24°C—rather more than optimists had expected and somewhat less than pessimists had feared. The report was discussed by the former residents of the Maldives and the Bay of Bengal in the various places to which they had fled, by the refugees from the Great Italian Desert and the California salt-marshes, and a few lucky predictors raised their glasses of Hebridean wine. It was an occasion for many of the world's two billion displaced people to reflect on the billion-plus members of their former homelands who had succumbed to floods, hurricanes, droughts, starvation, and the ravages of the six great pandemics that had swept the globe during the past four decades—and it was also the occasion for an unusually popular podcast (a quaintly anachronistic form of communication that survived only in backwater academic circles).

Nobody would have expected an obscure professor of classical phi-losophy—one of the very few remaining members of that apparently useless specialty—to engage so large an audience, but the piece struck a chord. Of course, the world's discussion fora had been filled for decades with recriminations against the voters of the early twenty-first century nations who had, time after time, elected politicians who vowed that anthropogenic global warming was a hoax, or, later, that the problem could be addressed without diverging from established patterns of energy usage. Especially in the disaster zones and in the refugee camps, survivors had angrily debated the sources of the appalling irresponsibility that had caused the climate crisis. The professor did not address that question, but she did point out, convincingly, that some such catastrophe had been pre-dicted by an ancient thinker.

Even the opening of the podcast was quaint—it began with the now defunct subjunctive:

> Were Plato living today, he would be hard pressed to avoid saying "I told you so."[2] Although he did not think of democracy as we do, and despite the fact that he would have seen our political systems as pecu-liar types of oligarchy, one of his central messages was that government is too difficult to be left to the ignorant. If those who do not know important things are allowed to have a say—*any* say—in the crafting of policies, there is a serious danger that foolish things will be done or that necessary things will be left undone. So-called democratic societies are disasters waiting to happen, and, in the early twenty-first century, the trouble finally erupted on a grand scale. Expert advice, even when loudly and forcefully given, was ignored. Sacrifices that were urgently needed were neglected. Our devotion to a misguided political system led to the irreversible modification of the one planet we have—and to the deaths of well over a billion people (not to mention the reduction of billions more to lives of hardship and squalor).

The professor continued with some scholarly details, but that opening paragraph, for all its stylistic infelicities and its archaic usages, resounded in the consciousness of a broad public. At last it was clear what had gone wrong. Our species had paid the price for its addiction to a political fetish, its obsession with democracy.

II

It may not go as badly as my story suggests, or it may go worse. According to the expert consensus on climate science, the best we can hope for, if nothing is done, is a rise of 2–3°C in the earth's mean temperature by century's end.[3] A really pessimistic assessment would predict 6–7°C. My imagined figure is roughly in the middle of the interval. Even if the rise were at the low end of the range, rising sea levels would eliminate low-lying islands and coastal areas—like the Maldives and the Bay of Bengal. The Italian desert is more of a conjecture, and, even with an increase of 5°C, the *bella campagna* might avoid that dreadful fate, if there were some (improbable) shift in weather patterns. That increase would almost certainly inundate California's central valley, producing a large body of salt-water broken by patches of marshland (and putting an end to California agriculture). Hebridean wine is a whimsical possibility, but the traditional homelands of famous vintages would almost certainly be unable to grow the necessary grapes. Increasing frequency and severity of destructive weather events—droughts, floods, hurricanes—is a straight-forward effect of rising average temperatures. Rapidly melting snowpacks will cause summers marked by initial floods and later droughts; floods will lead to contamination of water supplies. Shortages of water, pollution of springs and rivers, disruption of agriculture, and mass migrations will generate huge outbreaks of disease and even allow for the rapid evolution of new disease vectors.

During the past three decades, an overwhelming consensus has developed among the world's climate scientists. As James Hansen, Steven Schneider, and Michael Mann have documented, a plot of the earth's mean temperature against time shows extraordinary correlation with the levels of greenhouse gases (primarily but by no means exclusively carbon dioxide) in the atmosphere.[4] Correlation is not proof of direct causation, as every scientist knows and as climate deniers repeatedly insist, but, in this instance, not only is there an exquisite correlation, but also a mechanism—recognized by earth and atmospheric scientists for well over a century—for the increase in temperature when levels of carbon dioxide rise. Further, there is no plausible alternative mechanism for explaining the effect. When your favorite fictional detective assembles a dozen or so correlations between the murder and a suspect, when he shows how the suspect could have done the dreadful deed, and when there is no plausible rival account, his fictitious less gifted rivals and his actual devoted readers do not withhold assent. When climate scientists offer a sample of clues larger by several orders of magnitude, when they can demonstrate the potential mechanism, and when they can decisively eliminate proposed rivals, it would be stubborn folly not to accept their claims.

The minimal consensus estimate is that, *even were we to act immediately to control the emission of greenhouse gases*, the earth's mean temperature would rise by 2–3°C by 2100. Charitably, and perhaps too optimistically, I've taken that range as the base line, even if we continue to do nothing. Climate scientists offer different models of the factors affecting the mean temperature (since there is disagreement about the weight to be assigned to factors that might force or dampen the greenhouse mechanism), and they diverge in their assessments of rates of sea-level rise, and the effects on weather patterns and on different geographical regions. Some effects are, however, uncontroversial. Nobody doubts that the mean sea level will rise by enough to flood certain low-lying places. Nobody questions the fact that extreme weather events are increasing in frequency, that the distribution of weather shifts so that what was previously rare becomes more common and that the new "rare" events (the new tails of the distribution) are unprecedented.

From this state of consensus arises a multi-tiered debate. The ground level concerns the existence of anthropogenic global warming, and here the shouting should long have been over. The second level addresses the question of whether there will be unfortunate, or catastrophic, consequences of likely increases in the earth's mean temperature. *Ideally* that level of debate would proceed by assigning precise probabilities to the intervals within which the increase might lie—for example, by declaring that the probability of an increase of 4–5°C, given no steps to curb the emission of greenhouse gases, is 0.19—and by detailing the exact consequences of an increase of the type—say by delineating the regions of the world that would be submerged, identifying which would be affected by severe drought, and so forth. *Realistically* nothing of that sort is possible. Even if you restrict attention to the consequences for human well-being— a sensible restriction, given the difficulties of considering all the ecological variables that might affect the survival of a nonhuman species and the wide variation in human concern about species conservation—gauging sea-level rise is already difficult, and attending to the host of factors that might contribute to drought, disruption of agriculture, frequency of severe weather events, and spread of disease would be vastly harder. Any sensible debate at this second level would be better framed by specifying some notion of "catastrophic consequence" and asking, with respect to particular ranges of temperature increase, what the chances would be of avoiding that outcome: thus, you might declare that a situation in which a billion human lives were lost over a period of a decade, as a result of global warming was an unacceptably catastrophic consequence and ask for the probability of *avoiding* this outcome, given a rise in mean temperature of 3.5 to 4.5°C. Although any answer to that question would be a matter of expert judgment, it seems likely that those who have immersed themselves in articulating the wide variety of scenarios that could affect human well-being, and human survival, through the variety of factors mentioned—rising sea-levels, drought, water-contamination, disruption of agriculture, modified patterns of disease transmission, forced human migrations—would suppose it very improbable that the human losses in the later decades of the twenty-first century could be held to a billion if the earth's mean temperature were to rise by 3.5 to 4.5 degrees. Deriding expert judgment of this sort is no more reasonable than ridiculing the planning of prudent parents who, after thorough reflection on their circumstances, their prospects, and the needs of their children, decide on various insurance and investment strategies.

The third level embeds the conclusion of the second level in as thorough an examination of the likely outcome of alternative strategies as we can achieve. The reasonable core of the protest against the existing proposals to restrict the emission of greenhouse gases consists in the correct observation that these proposals impose burdens on people now living and on our descendants. Comparative analysis is critical. You need to understand not only that the consequences of unchecked behavior are unacceptably bad, but that they are worse than those that would result from applying the brakes. If, as I believe, the debate ought to be over at the ground level and the contours of a resolution are already visible at the second level, the third-level issues need far more focused articulation than they have so far received. Our political discussions urgently require a thorough survey of as many alternatives to business-as-usual as informed and imaginative thought can supply, and as careful and precise a delineation of the human consequences as can be achieved. If expert judgment enters into the second level, it does so more crucially here, for the selection of particular scenarios as likely is itself a matter of weighing the strength of factors that defy any ability to provide numerical estimates, and the further enterprise of comparing the chances of their occurring would involve the kinds of judgments about which even the most knowledgeable people justifiably feel queasy. Perhaps we can hope that the exploration of alternatives would be sufficiently imaginative to unearth some plausible scenario that would clearly be superior to the most prominent rivals—as might occur, for example, if there were some clever and convincing suggestion for harnessing sources of energy that would pose no environmental problems. In the absence of that, it may be necessary for our species, in full self-consciousness of what is being done, to take a gamble. We may find ourselves in the predicament William James conjures up at the close of "The Will to Believe": caught in a blinding blizzard on a high pass, the traveler must strike out in *some* direction, even though he is fully aware that he may be heading toward a precipice.

The fourth and final level brings to the deliberations of the third some important constraints from ethics and the theory of justice. The burdens and benefits of particular ways of responding to the rising mean temperature of our planet fall differently on different groups of people, on those who live in some places rather than in others, on members of societies who have gained from previous use of carbon-based fuels as compared with those whose nations are in the process of industrializing, and on those who live now and those who will come later. Even if the consequences of the array of possible responses developed at the third level are entirely clear, it would still be necessary to subject those responses to ethical scrutiny, to ensure that all those affected—and that includes all members of our species from now into the future—are treated fairly. Despite skeptical concerns that fundamental ethical issues are incapable of resolution, global warming presents us with questions about justice that human beings have to answer—and that we have to answer together.

I have only sketched the debate that nature demands of us, without attending to all sorts of considerations that would have to figure in any detailed outline. For present purposes, however, a sketch is enough. For my aim is to understand how our current misconceptions about democracy, and about what a commitment to democracy requires of us, interfere with the global political discussions we so urgently need.

III

In countries all over the world, political attention to climate change has waned. Even nations that appeared committed to imposing serious restrictions on the emission of greenhouse gases are retreating from what their manufacturing entrepreneurs view as "intrusive regulation," and in Britain and the United States the situation is particularly dire. One obvious factor in accounting for the retrenchment is the state of the global economy: essentially, nation-states are leaping to a conclusion about the third- and fourth-level debates because they see their current economic prospects as endangered. Yet the turn away from any active pursuit of a prudent climate policy is reinforced by public skepticism about the need for attention to the issue. In the wake of the disclosures about emails among climate scientists, and about a (minor) error in a section of the IPCC report, many media sources have trumpeted the idea

that "global warming is a hoax." A sufficient number of American senators and members of the House of Representatives are disposed to accept
that view to render serious policymaking by the world's largest economy
impossible. Their recalcitrance was already problematic during a period
when the United States was led by an intelligent and far-sighted president who, toward the end of his time in office, succeeded in coaxing the
world to accept commitments to reduce their emissions of greenhouse
gases. Although the targets agreed on in the Paris accord of December
2015 fell considerably short of what is urgently needed, they represented
a clear step in the right direction. Since the election of November 2016,
the situation has worsened considerably. The Trump administration, with
its absurd dismissals of scientific evidence, seems gleefully bent on accelerating the transition to climate disaster.

Polls and surveys are sometimes viewed as offering a more comforting
picture. The most recent survey from the Yale Program on Climate
Change Communication reports that the percentage of American citizens
who now believe that global warming is occurring is 70 percent, with 49
percent of Americans being "extremely sure" about this.[5] The percentage
who attribute global warming to human activities is now 58 percent.
Some 62 percent are at least "somewhat worried" about the effects of
global warming, but only 21 percent are "very worried" (although this is a
sharp increase from the roughly 10 percent who confessed to being "very
worried" in 2015). Apparently, though, the higher levels of anxiety stem
from reactions to recent weather patterns and extreme events, rather than
to the judgments of climate scientists. Only about 15 percent understand
that almost all climate scientists have "concluded that human-caused
global warming is happening."

The beliefs of citizens continue to lag behind expert knowledge, and
consequently underestimate the future threats. The result is continued tolerance of climate inaction. The inaction of successive US administrations
thus reflects the popular mood. Moreover, one might think, that is just as
it should be in a democracy. For, on a simple conception of what democracy entails, policy should reflect the will of the public. If the public does
not think a particular issue should be addressed, then it is entirely right

that nothing should be done about it. Plato saw this as a fundamental commitment of democracy, and, because he understood that people may be massively deceived—or misled—about what is in their interests, he drew the conclusion that democracy is a political disaster.

Many friends of democracy agree with Plato about the commitment of the political system they champion. Since they must also concede the possibility that the public can be seriously mistaken in its factual beliefs— and recognize that possibility as realized on various occasions in various societies—they are forced to Plato's intermediate conclusion, to wit that democracies sometimes (perhaps frequently) pursue policies that are contrary to the interests of a majority of their citizens. If this is not to count as a disaster, that must either be because the mismatch between policy and interests occurs less frequently under democratic regimes, or because there must be some special value in the kind of freedom democracy allows that counterbalances the thwarting of broadly shared aspirations.

It is possible to conceive ideal political systems that respond perfectly to the interests of the majority of citizens: Plato himself offered one utopia, in which the perfect match was supposedly guaranteed. One part of the defense of democracy consists in denying that anything close to complete alignment of policy with majority interests can be achieved in any world in which human beings run the show: no system of education could produce the guardians with their combination of deep insight and incorruptibility. Yet even if you suspended your doubts about the inevitable venality of political leaders, there would remain a further advantage of democratic decision-making, a benefit accruing from the freedom democracy allows. There is something to the idea that, even though others might make better decisions on our behalf, we gain by acting for ourselves: the extra mistakes are outweighed by the freedom of self-expression.

How far can that go? Suppose there were a deep desire shared by most members of a polity, even common to the overwhelming majority of human beings. Suppose that desire to be so central to the lives and personal projects of people that, were it not to be realizable, those lives and projects would be importantly undermined and defeated. Would a democracy in which policies that put the desire at risk—or even ran contrary to it—be preferable to an autocracy whose policies promoted the realization of that desire? Is the freedom to choose for oneself so valuable as to override the contravention of most people's central aims and goals?

For many human beings, our hopes for those who come after us are entwined with our sense of what we are. Those hopes are expressed in the sacrifices we make for children and grandchildren, but they range more broadly than that. Our lives are diminished if we come to understand that we have failed to do what is necessary to allow our descendants, both those who have a biological connection to us and those who live in the communities in which we have resided, to have the possibility of leading valuable lives of their own. If we were to think that we had put their habitat at risk, done things that made it impossible for them to live securely and in health on our planet, we would view ourselves as having failed to discharge an important obligation to them. It would be little consolation to emphasize the supposedly great value of having participated in radically uninformed democratic decisions.

The alleged "fundamental commitment" of democracy is part of an overly simple picture of what democracy is: Plato and my envisaged friends of democracy are wrong. To be sure, democratic freedom is a valuable thing, but its value is grounded in the fostering of the interests of the citizens. Our votes matter because they allow us to make our interests apparent. To the extent that the preferences we express run contrary to our interests, those votes are meaningless.

Democracy is a historically evolving political system. Its popularity rests not on the naive thought that "government by the people" will result in policies that express the popular will, but on the clear capacity of democracy to solve the *problem of identifiable oppression*. That problem arises in societies where the overwhelming majority of the people can see very clearly that those who rule are acting in ways directly contrary to their interests: the deprivations and sufferings they experience can be easily and correctly traced to the actions of their oppressors. Enshrining particular limits in a constitution, whether written or unwritten, serves as a first line of defense against tyrannical predation, but if the rulers overstep the boundaries, the machinery of elections and voting allows for their recall. Reflection on the last century, and on the dictatorships that have flourished in many regions of the globe, makes clear the importance of finding a solution to this problem.

Contemporary human societies live, however, with a different difficulty: the *problem of unidentifiable oppression*. Even in the most affluent contemporary democracies, some citizens suffer from governmental policies that deprive them of opportunities to pursue their legitimate interests. Neither the children who go to ill-staffed, poorly equipped, dangerous schools, nor those who care for them (often a single parent or a weary grandmother) can easily recognize the sources of the conditions that so tightly cramp their future prospects from an early age. The effects of a diffuse cloud of governmental and bureaucratic decisions "trickle down" (for once the phrase is apt) into their soggy lives. Only when democratic government is reinforced with public knowledge can democracy cope with problems of this sort. Widespread ignorance of the factors that thwart the citizens' interests makes "government by the people" an empty piety. Plato was partly right, after all.

In a world with a complex division of labor, in which consequential decisions turn on pieces of specialized knowledge whose *formulations* are typically incomprehensible to the vast majority of people, problems of unidentifiable oppression are everywhere. If democracy is to address many of the sources of current oppression, it is not enough for it to stop with the defense against evident tyranny (important though that defense is). Democracy needs to evolve further.

IV

At this point, an obvious thought is likely to surface. Democratic societies are not only committed to constitutional guarantees and opportunities to recall those who overstep them, but are also marked by "free and open" discussion. This supplies precisely what is needed. Equipped with their native wit, common sense, natural rationality (or what you will), the citizens are able to listen to and appraise the issues that arise for them, to come to recognize the things they need to know. Contemporary enthusiasm for the magic wrought by "fair and balanced" debate—matched only by ardor for the magic worked by the "free market"—translates into the modern context points that were elegantly and eloquently made centuries ago. John Milton asks, rhetorically, "Who ever knew Truth put to the worse in a free and open encounter?" and John Stuart Mill celebrates the "clearer and livelier perception of truth produced by its collision with error."[6] Perhaps Milton and Mill were right to be convinced that public debate would produce such benefits in the controversies of most concern to them. But that is not the way we live now.

Milton, Mill, and their successors presuppose a principle of *evidential harmony*: public discussion of controversial issues occurs in such a way that those who make the final assessments are able to do so on the basis of the evidential support accruing to the rival positions. That principle is plainly flouted in the debate with which this essay began: only a tiny number of those whose interests will be affected by climate change are in any position to recognize the evidence for or against any of the principal claims that figure in the public controversy. To suppose that the newspaper reports they read, the radio broadcasts they hear, the television news they watch, or the internet sites they visit provide them with any "clearer and livelier perception of the truth" is an absurd fantasy. Attention to the full range of these media sources makes it evident that there are many devices available to those who wish to put Truth to the worse. Climate deniers are handsomely funded by industrial sugar daddies, and able to produce slick videos to market their message. They appropriate the rhetoric of the virtues of free discussion, praising a worthy ideal while simultaneously creating conditions—arenas in which production values substitute for reasons—that foil any chance of its realization.

In the contemporary world, market mechanisms are unlikely to favor the dissemination of accurate information. Even if inaccuracies are detected, and even if they are deemed important, their promulgation can be explained away. Media sources can expect large profits by substituting sensationalism for sober reporting, entertainment for enlightenment. Especially in societies where there are deep political and religious differences, market niches can be defined in terms of a constellation of attitudes to be continually reinforced by the latest "news": those who tune in can be sure that their fundamental convictions will not be perturbed. Moreover, the less accurate a picture of the world the citizens have, the less able they are to engage in critical reflection on their favored sources of "information." The world they inhabit is remote from those envisaged by Milton and Mill.

Yet, even if these familiar features of the transmission of "public knowledge" were absent, it would still be hard to cope with the problems of unidentifiable oppression we face. For an important point, descending from Thomas Kuhn—and ultimately from Ludwig Fleck—casts doubt on any possibility of timely resolution of technical controversies in a public forum. Kuhn's historical examples show clearly that the considerations underlying many scientific debates are subtle and delicate, that resolutions come slowly, and that a complex web of argumentation must be elaborated before anything close to consensus can be reached.[7] Scientific decision-making is not algorithmic (although philosophers continue to dream of making it so), nor is it the case that some single piece of evidence or argument convinces everyone. Hence, even given universal good will and impeccable honesty, the task of settling the climate-change controversy in public would pose enormous challenges. It is simply unrealistic to expect that democracy's commitment to "free and open discussion" will solve the many problems of unidentifiable oppression faced by the citizens of modern societies, or that open debate will create pressure on political leaders to craft policies answering to a deep and central interest of almost all people. Sustaining democracy seems incompatible with saving our planet.

V

Is this Plato's revenge? I hope not, and my hopes are founded in the thought that democracy is—must be—an evolving political system. Plato rightly linked education to his political system, and so too did one of democracy's premier champions, John Dewey. Education, serious education, has to be central to the democracies of the future (assuming that they continue to exist), but, as Dewey saw, democratization can itself be a form of education.[8] Beyond the superficial machinery of votes and elections, beyond the "free discussions" in which partisans of different views campaign for the allegiance of citizens, lies the engagement of citizen with citizen, the sharing of points of view, the mutual recognition. Reforming the educational systems of the contemporary world is probably an important step, but it will be too slow to address some of the challenges our species now faces: even if the voters of the next generations would have a firmer grasp of the issues pertinent to climate change, we cannot wait that long. In the short term, what is urgently needed is the restoration of trust in genuine expertise. How might that happen?

Through the refashioning of a type of democratic discussion that has often been central to flourishing government by the people. Citizens with different perspectives can be brought together, they can be put into conversations with representatives of pertinent areas of technical inquiry and encouraged to arrive at a consensus. Their testimony can then help a broader class of citizens to share the expert understanding: trust can be restored as small representative groups are "led behind the scenes," emerging to spread their findings to larger populations. This is an extension of the ideal of Well-Ordered Science, an ideal of scientific research as promoting social health and individual well-being.[9]

In recent decades, the authority of scientific experts has been radically eroded.[10] That is not because of some particular philosophical movement—postmodernism has not been introduced into the drinking water. Nor is it because the findings of scientists have been decisively shown to be unreliable. Rather, the trouble stems in part from the need for scientists to pronounce on urgent issues before there is a clearly articulated consensus that can be presented and defended in simple terms, in part from the workings of the "credit economy" that underlies the practices of scientific communities, and in part from the inevitable entanglement of scientific work with judgments of value.

We inhabit a world in which there are many large problems with respect to which scientific pronouncements must be tentative, and discussion of such problems frequently involves public debate among "experts." Further, the welcome diversity of scientific perspectives is often encouraged by competition among scientists who vie for credit. Even when a consensus seems imminent, those with divergent views have incentives to take their ideas to a lay public that is in no position to judge their merits.

Most fundamental, however, is the increasing recognition that, in choosing their problems, in conducting their research, and even in validating their conclusions, scientists have to decide that the hypotheses they entertain and the practical tools they forge are well-established enough for the purposes to which they will be put. Value judgments enter here. Once that fact is appreciated, the scientists who put forward a consensus view can be charged with deserting the scientific ideal, of succumbing to illicit prejudices. The situation is further complicated by a widespread distrust of any possibility of serious discussion of values. The thoughtful climate scientists who make judgments about human consequences are effectively assimilated into those who announce views that harmonize strikingly with the aims of the private corporations that fund their research.

Within the family and in small communities, profitable discussion of values and aims is possible, even common. The right worry about "experts" is not that they are importing values—that is inevitable—but that they introduce values that would not be sustainable in a broadly democratic discussion: they promote special interests, not the deeply shared concerns of most members of our species. We urgently need channels of communication that will allay that kind of doubt. Hence my (tentative) suggestion that small representative groups of citizens go behind the scenes, returning to report on the nature of the consensus and the values that are in play. To answer Plato's challenge, something of that sort must remedy the public ignorance to which our children's future is currently held hostage.

In the end, the democratic conversation must become truly global. With respect to climate change, the characteristic procedures introduced and refined by the IPCC need to be embedded in discussions in which citizens from all parts of the world, and from diverse perspectives, come together to share in decision-making. The town meetings Tocqueville rightly saw as so vibrant a part of American democracy might take place on a broad scale, as citizens—citizens of the world—join to carry on the multilevel conversation we discussed earlier. Nobody should suppose that this will be easy, or even assume that it will necessarily succeed, but if Plato is not to have the last mordant laugh, it is hard to think of any serious alternative.

CHAPTER 12

DEMOCRACY AND THE PROBLEM OF PSEUDOSCIENCE

MICHAEL RUSE

I take it as nigh axiomatic that the sign of a healthy democracy is that members are given as much freedom as possible. No one is arguing for absolute freedom. Men should not be allowed to go out and rape at will, for instance. But freedom where there is no good reason to deny it. If you want to go to a Catholic church, that is your business. If you want to go to a Protestant church, your business again. And if you want to go to no church at all, still your business. There are going to be debates about where to draw the line. No one should be allowed, without reason, to yell "fire" in a crowded cinema. No one should be barred from saying that they don't much care for the current crop of leaders. What about getting up in a public square and saying, "Hitler was right about the Jews. Pity there weren't more gas ovens"? In the United States, you are allowed to do this. In Canada, you are not. Different views, with reasons for or against either.

WHAT IS PSEUDOSCIENCE?

What about pseudoscience? This is a slippery notion, a bit like religion, easier to spot than to define.[1] Roughly, though, I take it that pseudoscience is a body of knowledge or of claims purporting to be knowledge. This is knowledge or purported knowledge about the physical world, in the same sense that regular science is. However, it is driven by cultural values, broadly construed—religious, philosophical, political, and more—rather than by the empirical world. Moreover, when its claims come into conflict with the physical world, it is the physical world that has to concede ground rather than the body of claims. The judge of what is to be labeled a pseudoscience is the professional scientific community, and often cries of "pseudoscience" are made most often and sound most shrill when for one reason or another the professional science is under attack.

A paradigmatic example of a pseudoscience is the eighteenth-century belief system (and practice) of mesmerism. Named after the German doctor Franz Mesmer, it postulated the existence of vital forces in living beings that were open to discovery and manipulation—in other words, a kind of "animal magnetism"—and that its practitioners regarded as a way to cure many illnesses and diseases. It was very popular—it is almost a mark of a pseudoscience that it is popular—and from the first was criticized and condemned by actual scientists. In 1784, the French king Louis XVI set up a commission—including Benjamin Franklin and Antoine Lavoisier—to study mesmerism, and they drew conclusions that were very critical. Not that their report had much effect, because mesmerism rode high until the middle of the nineteenth century, when it ran out of steam.

Although these things are usually portrayed as black or white, they often are not. Evolutionary thinking is a good case in point. A good case can be made for saying that, until Charles Darwin published his *Origin of Species* in 1859, evolutionary thinking was more or less—usually more—a pseudoscience.[2] It had little hard empirical evidence and was driven by values—in particular, the ideology of progress. Things getting better in the social and cultural world—better healthcare and so forth—reflected into the belief that things are getting better in the living world. From the blob to the human, or (as people used to say) from the monad to the man. Charles Darwin's grandfather Erasmus Darwin is a case in point. He was an ardent evolutionist given to expressing his ideas in verse:

Organic Life beneath the shoreless waves
Was born and nurs'd in Ocean's pearly caves;
First forms minute, unseen by spheric glass,
Move on the mud, or pierce the watery mass;
These, as successive generations bloom,
New powers acquire, and larger limbs assume;
Whence countless groups of vegetation spring,
And breathing realms of fin, and feet, and wing.

Thus the tall Oak, the giant of the wood,
Which bears Britannia's thunders on the flood;
The Whale, unmeasured monster of the main,
The lordly Lion, monarch of the plain,
The Eagle soaring in the realms of air,
Whose eye undazzled drinks the solar glare,
Imperious man, who rules the bestial crowd,
Of language, reason, and reflection proud,
With brow erect who scorns this earthy sod,
And styles himself the image of his God;
Arose from rudiments of form and sense,
An embryon point, or microscopic ens![3]

Erasmus Darwin explicitly tied his biology to his philosophy. This idea of organic progressive evolution "is analogous to the improving excellence observable in every part of the creation; such as ... the progressive increase of the wisdom and happiness of its inhabitants."[4]

The term "pseudoscience" (or "pseudo science") only came into being and use in the nineteenth century, but people knew a pseudoscience when they saw it. Noted already is how professional scientists are usually irate to the point of being shrill about pseudoscience, especially when they see it as threatening their own interests. In Britain, even in the middle of the nineteenth century, many of the scientists were Anglican clergymen supported in their work—especially if they worked at Oxford or Cambridge—by their church.[5] These people—men like the professor of geology at the University of Cambridge, Adam Sedgwick—were not crude literalists. They knew the earth was old and they didn't believe in a universal deluge. But they saw the naturalistic theory of evolution as a threat to their status and beliefs and reacted with fury. To a mid-century evolutionary tract, the *Vestiges of the Natural History of Creation*, Sedgwick responded first with an eighty-five-page diatribe in the leading journal, *The Quarterly Review*, and then, in a new edition of a little sermon (thirty-five pages on the proper conduct of students at university), he added a five-hundred-page anti-*Vestiges* preface, and for good measure a three-hundred-page epilogue.

Things were turned around completely when Darwin published his theory of evolution through natural selection; but, at the time, I am not sure that Sedgwick's response was so very wrong. Neither Erasmus Darwin nor the anonymous author of *Vestiges*—known now to be the Scottish publisher Robert Chambers—had given much empirical evidence for their beliefs. In Chambers's case, particularly, there were all sorts of bogus arguments, for instance that the patterns of frost on windows—"frost ferns"—suggest the spontaneous generation of life. It was clear that everything was fueled by enthusiasm for the idea of progress—something that runs directly counter to the central Christian doctrine of Providence—and by the ideology of making things better for ourselves as opposed to putting everything in the hands of God.

PSEUDOSCIENCE TODAY

Now what about pseudoscience today and its relationship to democracy? Pseudoscience is certainly alive and well, to use a somewhat unfortunate metaphor. Anti–global warming, anti-vaccination (anti-vax), anti–genetically modified foods, for a start. Some of these systems are clear-cut examples of pseudoscience—such as anti–global warming. Some are contested somewhat. Chiropractic might be a case in point. Some swear by it. Others think it phony—a pseudoscience—especially when it makes grandiose claims about cancer and so forth. Amusingly, several years ago, the then very new Faculty of Medicine at Florida State University was offered by the legislature a large sum of money to start a department of chiropractic medicine. Immediately, the rather insecure school raised the cry of "pseu-doscience" and the offer was rejected. This despite the fact that before and after the incident the local medical establishment would often refer patients to chiropractors. The new school just could not afford to be connected to such a flaky business, especially with the grandiose claims about curing cancer. Harvard might get away with it, but not Florida State.[6]

What about democracy? However strongly you may feel that some-thing is a pseudoscience, you are hardly going to stop people believing in it. At least, not in the sense of prohibiting it. To be honest, I am not quite sure how you would prohibit it. I happen to think that opposition to global warming theory is bad—really, really bad—but I am not sure that, through some kind of mind control, I want to make even the president of the United States of America agree that global warming is real and a threat. Mind control like you get in *1984*, although I guess that today, instead of rats, you use Fox News. One can, however, think of obvious examples where you might want to block the dissemination of pseudoscientific ideas. Suppose that someone had the view that sexual intercourse with young children is the secret to physical health. Even worse, suppose they think that drinking the blood of young children adds to one's life-span. In cases like these, we don't want people getting up and being allowed to promote their beliefs on television. Are they more like crying fire in a cinema or like saying Hitler was right on about the Jews? Most of us would say that, if there is any hint that such views might push people into action, then they are more like the cinema case and should be prohibited.

Much as I dislike uninformed criticisms of global warming—if any-thing is pseudoscience they are, with no empirical evidence and value-driven arguments (for instance about the need to keep jobs in the coal-mining industry)—I don't see them in the drinking-the-blood-of-young children category. At least, not in any direct sense. In this age of the internet, apart from the sheer impossibility of banning open discussions critical of global warming theory—although the Chinese do seem to be pretty good at this kind of thing—it seems to me to be inherently wrong in a democracy to stop people from disseminating their views, however crackpot they may be. I think the Mormons have some unbelievably daft ideas, but I don't want to ban the Church of Jesus Christ of the Latter-day Saints. Also, as the case of evolution makes clear, you need to let ideas thrive, however silly and annoying they may be. Evolution was a pseudo-science until the *Origin*, but one doubts that Darwin would have done what he did do without the background that this pseudoscience gave to him. More generally, I am Popperian in thinking that you should gen-erate hypotheses if you want to move forward. Or, as Mao said in 1957, launching a short-lived movement to let intellectuals offer open criticism of his regime, "Letting a hundred flowers blossom and a hundred schools of thought contend is the policy for promoting progress in the arts and the sciences and a flourishing socialist culture in our land."

A hundred flowers blossom? Yes, but then one needs to start cutting them down. I think anti-vaxxers are really dangerous. Their ill-formed and pernicious views put young children at serious risk. No child should go through life with the terrible eyesight of my father, a result of child-hood measles. Reluctantly, however, I would not stop fanatics from preaching the dangers of vaccination. At the same time, the obligation is on me to oppose the views, publicly, in an informed manner. Tolerance is not an excuse for doing nothing. I don't think one necessarily has to spend all of one's life systematically going after every pseudoscience in a row—although perhaps those of us who are philosophers have a greater obligation than most—but one should do something.

I am not Mother Teresa, but I have tried to follow my own urging, and I want now—providing a case study to flesh out what I have been talking of in general terms—to talk about my own encounter with pseu-doscience. How I was led into it, what I did, and what followed.[7]

CREATION SCIENCE

I am by training a philosopher of science. In the 1960s, looking for a thesis topic—I was studying in England, in the United States this would be "looking for a dissertation topic"—I, like a small number of others, realized that biology was a science crying out for philosophical attention.

I set about obliging.[8] Serendipitously, this was just the time when the philosophy of science community was feeling the full force of Thomas Kuhn's *The Structure of Scientific Revolutions* (1962), with the exhortation to turn to the history of science for understanding of today's science. Naturally, I turned to the Darwinian Revolution, the coming, in the mid-nineteenth century, of Darwin's theory of evolution through natural selection. Taking to heart what Kuhn told us, namely that we philosophers had to take the history of science as seriously as did the historians, I spent a sabbatical year at Cambridge attending lectures and seminars, as well as working in the Darwin Archives in the University Library.

This resulted in my writing and publishing, in 1979, *The Darwinian Revolution: Science Red in Tooth and Claw*. I joke that this is the book that I wish I could have read ten years previously when I first became interested in Darwin. Like all good jokes, it was not really a joke. I set out to give a comprehensive overview of the Darwinian Revolution, explaining what went on and why. Forty years and twenty-five-thousand copies (twice that in translation) later—still in print—I think I succeeded in my aim. What is important for my story here is that I came to the project as a philosopher and this showed. I was interested in such questions as the evidence for evolution and the nature of natural selection as a mechanism. Not to mention issues to do with science, religion, and the like.

All of this know-how suddenly found an immediate and unintended use. The moment when the book was published was precisely that when the American biblical literalists, the fundamentalists of old, now rechristened the creationists—a vocal subset of evangelical Protestantism—were stepping up the pressure.[9] That subset wanted to have its views—six days of creation, six thousand years ago, miraculous creation of organisms, including—especially including—Adam and Eve, and then a universal flood—included in science classrooms of the United States. Because of the First Amendment's separation of church and state, this could not be done directly. A subterfuge had to be created. This was simply to argue that all of the claims of the creationists were in fact backed by good science. Thus was born scientific creationism or creation science,[10] something designed expressly to get around the First Amendment.

Note that creation science is not religion; it is something designed to take on professional science. The religious person might, for instance, think that the story of Noah's flood is true, just not literally true. It is not something that justifies or was ever intended to justify a worldwide deluge. It is rather allegorical or metaphorical. In the story, after the flood is over, Noah gets drunk and one of his sons makes fun of this. The story truly is about the futility of simplistic solutions to complex problems— wipe them all out and sin will be gone. Things just are not that simple. One can hold to this and be perfectly comfortable with modern science.[11] Creation science denies that we are dealing with a metaphor. It uses all sorts of pseudo arguments to claim that the flood really did happen. In short, it is as much a paradigm of pseudoscientific thinking as was mesmerism two centuries earlier.

So there we were. On the one side was creation science, something invented to get around the First Amendment. Something that its supporters wanted to get into the science classrooms—note, science classrooms, not general studies or some such thing. On the other side was I, serendipitously preadapted to take on creation science, for I knew the history and the philosophy of what was going on. Things came to a head in 1981, when the state of Arkansas in its wisdom passed a bill mandating the "balanced treatment" of Darwinian evolutionary theory and creation science in the science classrooms of the publicly funded schools in the state. It was signed by the governor, a man as surprised at being elected as he was unfit for the job. (Sound familiar?!)[12] At once, the ACLU sprang into action, opposing the new law and lining up a list of expert witnesses—these included paleontologist Steven Jay Gould, geneticist Francisco J. Ayala, and Langdon Gilkey, the most famous Protestant theologian of his day. I was added, almost as an afterthought, as the historian and philosopher of science.

I say as an "afterthought," but that is not quite true. The ACLU thought long and hard about including a philosopher—things could so easily go wrong—and it was not until I was virtually on the stand that they decided to use me. As it happens, their fears were well grounded. The state, convinced that philosophy is just a flimflam game, grilled me nonstop for three hours. This was at least twice as long as anyone else.

However, the nerve of the ACLU paid off. The judge used my testimony as the key support of his decision to rule the new law unconstitutional. He picked up on my definition of science—showing that creation science is not science but something pseudo, put in place for a religious purpose.[13] He quoted me verbatim.

(1) It is guided by natural law;
(2) It has to be explanatory by reference to nature law;
(3) It is testable against the empirical world;
(4) Its conclusions are tentative, i.e., are not necessarily the final word; and
(5) It is falsifiable (testimony of Ruse and other science witnesses).[14]

So here was a case where a pseudoscience posed a real threat to education and where it was right and proper to oppose it. A thriving democracy depends on its inhabitants being properly educated and trained. If you turn your back on science in a case like this, it is the thin end of a massive wedge that makes it only too easy to move on from evolution to global warming and so forth.

THE STORY CONTINUES

It is worth noting how enthusiasts for the Left are as much into pseudoscience—anti-vaxxers and those who are anti-GM foods—as enthusiasts for the Right—who are against Darwinism and global warming. The move from anti-Darwin to anti-vax and anti-GM might not be direct, but it is interesting how much anti-Darwinism there is in the thinking of those on the Left. They hate what they think are the harsh social Darwinian implications of Darwin's natural selection, beginning as it does with the struggle for existence. This all becomes very clear as we continue with our story, for the defeat in Arkansas was far from the end of the story. Indeed, I am sure that there is today more creationism being taught in American schools—including those getting government funding—than there was forty years ago.

It was clear after Arkansas—and it was clear to many of the more far-sighted creationists before Arkansas—that a full-frontal assault on the problem was going to come to grief. The First Amendment's separation of church and state is a barrier not to be passed. A more subtle approach was needed. This came at the beginning of the last decade of the century, with *Darwin on Trial*, by Berkeley law professor Phillip Johnson.[15] Here, we are introduced to the idea of "Intelligent Design Theory." Explicit reference to the Bible is minimized to the point of nonbeing. Indeed, explicit reference to the Christian god is minimized to the point of nonbeing. However, an intelligence—more likely an Intelligence—is demanded to make the process of creation—which might plausibly be evolutionary—fully effective. Very much a case of nod nod, wink wink, because it is clear—as was discovered later through a cache of private emails—that it is creationism of one sort or another that is hiding behind the veils.[16]

In the last decade of the last century, a number of writers, including Catholic biochemist Michael Behe[17] and Protestant philosopher/mathematician William Dembski,[18] picked up on Intelligent Design Theory (IDT) and started to flesh it out.[19] Behe cited a number of biological phenomena, including the bacterial flagellum (the whip-like tail that some bacteria use for motion) and the blood-clotting cascade (a series of chemical reactions involved in the coagulation of blood). He argued that these phenomena cannot be given a natural explanation. Dembski turned to theory, arguing that naturalism leaves holes to be plugged by an external, designing intervention. Immediately critics descended, notably the Catholic biologist Kenneth Miller[20] and the Quaker philosopher Robert Pennock,[21] who showed that neither scientifically nor philosophically does IDT hold water. It is as pseudoscientific—and for many of the same reasons—as the original creation science. To note this connection, I refer to IDT as "creationism lite."

History repeated itself. Again, against the advice of far-sighted creationists or IDT enthusiasts, a court case was brought, in the town of Dover, Pennsylvania, this time, when the attempt was made to get IDT into state-supported classrooms. Again, the result was the same, with a quite conservative judge ruling firmly that IDT is religiously motivated pseudoscience, trying to slip around the First Amendment.[22] Two successes in court, but no ultimate victory. Finally, creationists decided to do what the wiser said they should have been doing all along. They now work at the local level, under the radar as it were. So-called charter schools are now very popular—and strongly supported by the Trump administration. These schools get public funds but they go their own merry ways when it comes to curriculum. It is hard to get statistics, because that is the very thing that charter school enthusiasts prefer not to disclose. However, given the enthusiasm that the evangelicals show for such schools, it does not require a very stable genius to infer that creationism is firmly ensconced in biology classes. This goes along with field trips to the Creationist Museum in Northern Kentucky.

As I have said, what is remarkable is the extent to which my fellow philosophers show enthusiasm for IDT, even those who claim to be atheists! They like the sense of meaning and of wholeness that you get out of this world picture, in contradistinction to the harshness of Darwinism.[23] Thomas Nagel is one who cozies up to people like Behe, while at the same time favoring some kind of strong, secular Aristotelian position. He tells us that we are presented with two options: blind law (which presumably means Darwinism) or "there are natural teleological laws governing the development of organization over time, in addition to laws of the familiar kind governing the behavior of the elements."[24] Nagel continues:

> This is a throwback to the Aristotelian conception of nature, banished from the scene at the birth of modern science. But I have been persuaded that the idea of teleological laws is coherent, and quite different from the intentions of a purposive being who produces the means to his ends by choice. In spite of the exclusion of teleology from contemporary science, it certainly shouldn't be ruled out a priori.[25]

It is worth noting that one of the biggest criticisms made by regular biologists of Jim Lovelock's[26] Gaia hypothesis—that the earth is a living organism—was that it was teleological, explaining in terms of ends rather than prior causes. This was a major reason many characterized it as a pseudoscience.[27] "Gaia—the Great Earth Mother! The planetary organism!"

John Postgate, leading British microbiologist and Fellow of the Royal Society, writes.[28] "Am I the only biologist to suffer a nasty twitch, a feeling of unreality, when the media invite me yet again to take her seriously?" He continues, writing that Gaia "has metamorphosed, in Lovelock's writings and those of others, first into a hypothesis, later into a theory, then into something terribly like a cult." Postgate's judgment: "It is pseudoscientific mythmaking." And his warning: "When Lovelock introduced her in 1972, Gaia was an amusing, fanciful name for a familiar concept; today he would have it be a theory, one which tells us that the Earth is a living organism. Will tomorrow bring hordes of militant Gaia activists enforcing some pseudoscientific idiocy on the community, crying 'There is no God but Gaia and Lovelock is her prophet'? All too easily."

TAKING STOCK

Let me start to gather together the threads of my discussion. Pseudoscience is a mockery of real science. It is value driven and has scant regard for the facts. Worse, it manipulates the facts to satisfy its value-impregnated view of the world. This said, there is no place in a healthy democracy for banning pseudoscience. Apart from the difficulty of doing this effectively, especially that now we are in the age of the internet, there are reasons for not so doing on what we might call Kantian grounds and on utilitarian grounds. If we are to respect people as people—treat them as ends in themselves and not as means—then this means letting people believe and promulgate daft ideas. There will be limits, of course, but these are limits, not the thin end of the wedge leading to total prohibition. Tolerance does not mean acceptance or indifference.

Pseudoscience is not just an insult to human rationality; it is or can be dangerous. This can be short-term: anti-vaxxers are already having success in places like California, with the consequent rise of hitherto-vanquished childhood illnesses like measles. This can be longer-term: people in India and elsewhere on the globe are starving because misguided fanatics in the West do all they can to prevent the development and use of genetically modified organisms. This can be short-term and long-term: already global warming is causing disruptions. Ask me, I live in Florida where we are experiencing one horrendous hurricane after another. I had just finished removing the detritus from last year, including a massive tree right across my garden and smashing against the windows in the living room, when hurricane season was on us again. Down the road, things are only going to get worse. This truly is crisis time, folks!

Therefore, as responsible citizens, we have an obligation to fight against pseudoscience, showing its faults and trying to limit its influence. This obligation is especially strong for those of us with the tools or weapons to enter the fight. My knowledge of the history of Darwinism is a case in point.[29] Although others might disagree, it is legitimate for the cobbler to stick to his last. Significantly, although my field of expertise is biology, I am probably most incensed by the anti-vax movement. I was raised in the 1940s, when vaccinations were just coming in and we children were starting to benefit. I still remember the horrors of polio and how so often in the summer swimming pools would be closed for fear of contamination. Then, when a teenager, my mother died and my father remarried to a German woman, whose family were deeply into the Rudolf Steiner movement, anthroposophy. My father became enthused to the extent of leaving his job and going to work in a Steiner-based school, a so-called Waldorf school. Steiner argued that childhood illnesses are an essential aspect of physical and spiritual development.

For over fifty years (when I go back to England from North America, where I have lived since 1962), I have listened to ongoing diatribes about the evils of vaccination. This from a father whose dreadful eyesight was a function of measles. However, I have stayed out of the anti-vax fight. I am not medically trained nor am I a statistician—the two attributes most needed in this particular battle. I urge from the sidelines! I have other work to do.[30] In fact, what I am doing now is part of that work. (You didn't think I was writing this piece for pleasure, did you?!) I am seventy-eight years old and still at it. I don't have to answer to anyone. (I have been a nonbeliever for over fifty years. I still take the Parable of the Talents—I have not been put on this earth with gifts just to please myself—as my categorical imperative.)

Am I and others of my ilk—Philip Kitcher, who wrote most eloquently against creationism,[31] Ken Miller and Robert Pennock, who have written equally eloquently against IDT—just kidding ourselves? I remember, many years ago, being on a line protesting racial inequality. My neighbor turned to me and said, "You know, I think these things are like prayer. They make us feel good, but I doubt they have much overall effect on the real world." Is this what is going on here? Does our intervention and efforts make any difference in the fight against pseud-science? It is naive to think that we intellectuals have total or even the biggest control over things. It is very clear, to stay with the evolution case for now—although the global warming issue cries out for similar discussion—that external factors, basically political, have huge effects. I have said already, I am sure truly, that huge amounts of creationism and IDT are being taught in America at the moment. Fortunately, other than at farcical excuses for education like Liberty University, this is not occurring at the level of higher education. Secondary education is another matter, and a major reason for this is that legislatures and administrations have made this possible. As of this writing, the current United States Secretary of Education is Betsy DeVos, a passionate advocate for private education and for charter schools. On a recent visit to my home city of Tallahassee, she ostentatiously ignored the public-school system and visited only private (religious) schools. She and President Donald Trump are responsible for much of this.

External factors have huge effects. I am not sure, though, that this means they have all the effects or even all the significant effects. In the case of evolution, imagine what would have happened had we lost in Arkansas and then later in (Dover) Pennsylvania. Creationism would be taught widely in public schools, and for those who think that this would not have a knock-on effect on higher education, let me tell you that I have a bridge in which I would like to interest you. I do think that my own efforts, along with the efforts of Stephen Jay Gould and Francisco Ayala and Langdon Gilkey and Kenneth Miller and Robert Pennock and other fellow laborers, have made a very big difference, if only by preventing a very big difference! Pseudoscience is a nasty threat to democracy. We must do what we can to contain it. It is an ongoing fight with as many losses as successes. However, we can make a difference. We owe it to ourselves and our fellow citizens to do our best. We need to keep working!

DARWINISM AS RELIGION

Taking note of my own exhortation, let me conclude this short essay by talking very briefly about how my own researches have led me. Specifically, let me make mention of how, in recent years, my own work has focused a great deal on Darwinism and whether that early pseudoscientific side to the evolutionary enterprise has entirely vanished. After the Arkansas trial was over (at the end of 1981), although I had to spend much time defending myself against the criticisms of my fellow philosophers, my main worry was about why exactly it is that Darwinism causes such angst in the breasts of evangelical Christians. I knew that, for all of the talk about taking the Bible absolutely literally, this could not be the whole story. No one takes literally the claims in Revelation about the Whore of Babylon. She is always the pope or the Catholic Church generally, or someone further east like Saladin or Osama Bin Laden. There has never been any doubt in my mind that the Whore of Babylon refers to my late headmaster.

I took note of what these critics themselves said, namely that their big objection is that Darwinian theory is itself a religion, a religion that rivals Christianity. For many years, I pooh-poohed this idea. There is—as I still very much believe—a fully functioning genuine science of evolutionary biology, with Darwin's natural selection as the core causal force.[32] It was, after all, to defend this idea that I was called down to Arkansas. Gradually, however, I came to see the truth in the charge. Alongside the genuine science, there is a body of claims that truly functions as a religion. I make no claims about a hierarchy, even though there are days when Richard Dawkins, as people charged Thomas Henry Huxley with many years ago, does somewhat resemble a high priest, or even a pope. If you prefer, speak of a secular religious perspective. Either way, there is truly something more than pure science that challenges Christianity.

I have continued to explore this insight in a number of books. In *The Evolution-Creation Struggle*, I analyzed matters in apocalyptic terms, arguing that creationists tend to premillennial thinking and Darwinians to postmillennial thinking.[33] In *Darwinism as Religion: What Literature Tells Us about Evolution*, I explore these insights through the writings of poets and novelists.[34] My next book, *The Problem of War: Darwinism, Christianity, and Their Battle to Understand Human Conflict*, uses war as a case study to explore the thinking of Christians and Darwinians on so important and fraught a topic.[35] Through the Christian adhesion to Providence and original sin and through the Darwinian adhesion to progress and the virtues of struggle, I show that the differences are properly described as religious.

Now, in a proposed book—*A Meaning to Life*, to be published by Oxford University Press in the spring of 2019—I want to pull back a little and ask some bigger questions. Agree that the Darwinian Revolution was a watershed in Christian-Science dealings. Agree that Darwinism was in important respects turned into a religious rival to Christianity. What next? Is the Darwinian, accepting fully Darwin's theory including its application to our own species, committed then to a religious perspective? Having given up one religion, Christianity, is one now committed to accepting another religion, Darwinism? Or, is there a third way, one that takes Darwinian theory as a true foundation but that does not thereby embrace a religious perspective?[36]

In another recent book, *On Purpose*,[37] I began to explore what (somewhat pretentiously) I call "Darwinian Existentialism," where I try to show how one can accept the metaphysical implications of Darwin's thinking without plunging into religion.[38] Or worse, pseudoscience, bound up (as in the early years) with thoughts of progress—cultural and biological—making the running. Another approach is demanded, one that is, in respects, significantly chillier, and yet in other respects is not just more honest but ennobling and comforting. In some ways, it is close to existentialism. Jean-Paul Sartre makes the point about the alienation from God:

Existentialism is not so much an atheism in the sense that it would exhaust itself attempting to demonstrate the nonexistence of God; rather, it affirms that even if God were to exist, it would make no difference—that is our point of view. It is not that we believe that God exists, but we think that the real problem is not one of his existence; what man needs is to rediscover himself and to comprehend that nothing can save him from himself, not even valid proof of the existence of God.[39]

Then Sartre follows by trying to explain what this means for humankind:

My atheist existentialism . . . declares that God does not exist, yet there is still a being in whom existence precedes essence, a being which exists before being defined by any concept, and this being is man or, as Heidegger puts it, human reality.[40]

That means that man first exists, encounters himself and emerges in the world, to be defined afterward. Thus, there is no human nature, since there is no God to conceive it. It is man who conceives himself, who propels himself toward existence. Man becomes nothing other than what is actually done, not what he will want to be.

No student of modern science is going to accept all of this. Even a half-baked knowledge of human biology shows that it just plain silly to say that there is no human nature.[41] Humans are bipedal and rational and warthogs are not. It is true that human nature is variable—although, apparently, genetically we are nothing like as variable as many species—but to distinguish humans from warthogs is not bad science or motivated by racism or sexism or any other ism. The claim is true. To take a more specific example, for all of John Locke's horrendous stories about the ways in which people have treated their children, it is part of human nature to be loving toward children and especially so to one's own children. Of course, culture is involved. Perhaps culture can override biology and some people really do geld their children to fatten them up before eating them. Nevertheless, biology is the foundation. It is genetic that we humans can speak and warthogs cannot. Then, we speak different languages because of culture.

Qualifications notwithstanding, this approach nevertheless says that Sartre is right. We start from where we are. It is just a matter of where we are. The Darwinian says no one is a blank slate—and one very much doubts that Sartre, the quintessential Frenchman, truly thought that, either. We start from where we are and have to create meaning in this unfeeling Darwinian world.[42] There is no help from an external good God nor is there help from an external, progressive, value-increasing world process. Given this prospect, here too we can and must work through the items that give Christians and humanists meaning—family, friends, society, and more.

There is much more I could say and, before long, I am very sure I will say! For now, I will draw to a conclusion. Pseudoscience can be attractive even to the more skeptical of us. It is so easy to see it in the thinking of others and not in ourselves. If we cherish democracy, as we should, we need to be eternally vigilant. I am not arguing against trying to find meaning in our lives. I am not even arguing against religion, although it is not for me. I am arguing against thinking something is scientific when truly it is only value-impregnated, wishful thinking.

CHAPTER 13

THE FREEDOM TO BELIEVE—OR NOT

THOMAS DE ZENGOTITA

About fifteen years ago, an old friend from my teenage years contacted me out of the blue and we arranged to meet for coffee, talk about old times, catch up on the new, and so on. We hadn't seen each other for forty years and, to add to the sense of the distance we were bridging, the circumstances of our original acquaintance were specifically unusual. We were part of a crew of teenagers who hung around a huge navy base in the Far East—I was the son of a diplomat and most of them, including John, were sons of enlisted men, but the camaraderie that came with being part of a semi-delinquent "gang" of sixteen-year-old boys in the late '50s, thrown together in this enclave in a foreign land, meant that class distinctions were obscured and solidarity was strong.

And it held. In spite of the dramatic differences in the trajectories of our lives since—John's a retired cop and I'm an adjunct professor at NYU, contributing editor at *Harper's Magazine*—in spite of that, we fell easily into the kind of bantering companionship we had known as boys.

The second time we got together—for lunch on Long Island, a plan to go fishing after—he asked casually how I was doing as we sat down at the table and, taking a bit of a risk, I said I was actually kind of depressed. He asked why and I said (this was the beginning of March, 2003):

> Well, I'm afraid this Iraq thing is going to happen, he (meaning Bush) is just going to do it and it is going to be a disaster for everyone, especially the poor slobs who signed up for this believing that it's payback for 9/11.

I was thinking—I'll test the waters, I wonder what his politics are (he had said, in passing, some insightful things about crime and prisons the last time we talked)—but John wasn't going there. He paused for a long moment, looking quite baffled; finally, he said, as if in disbelief, "You're depressed about—Iraq?"

I missed the nuance and started to go on about the politics of it, when he interrupted and said, in an apparently wondering tone, as if trying to grasp something beyond his ken (but maybe not really), "I can see getting depressed if my kid got really sick or I got fired or my wife left me—but you're depressed about Iraq?"

Got it. His tone and expression remained puzzled, groping—but I caught on to the underlying attitude, the question he really had, the one he was too courteous to put into words. What he was really wondering was, "Who the hell do you think you are?"

I backtracked a bit too quickly—"Oh, of course, I don't mean that kind of depressed, I mean more like discouraged—like these people never learn, they keep sending our guys into these no-win situations (I knew he had fought in Vietnam) for bogus reasons . . ." But his point had been made, and I was conceding it in my manner. But, once again, courtesy and solidarity prevailed, and he moved smoothly on to bragging about his golf game—a sport he had just taken up last year and already he was shooting in the nineties.

Obviously much of today's skepticism about science reflects a more general resentment of "elites" that's abroad in Trump's America, but well-educated progressives who want to understand this phenomenon need to take a step back and consider some other, less obvious, factors. Least apparent, but perhaps most important, is this: feeling responsible for the fate of the world is profoundly unnatural—in some humanistic/ phenomenological sense of the word "natural." Most people, leading the lives they were thrown into, just don't feel that responsibility (unless aroused by some religious influence). No one blames a peasant in some mountain village in the Philippines if he's indifferent to his carbon footprint. Of course, the first thought on that score is that you can't blame someone who not only doesn't know about such things—but, realistically, *couldn't be expected to know*. For the well-educated progressive, the problem starts with exactly that—education and, in particular, with people who are in a position to know about such things without a massive investment of time and energy. They *choose* not to. Willful ignorance—that's the most immediate issue, that's what justifies our attitude toward science deniers in this society.

And *that's* because, to those of us who do feel that responsibility, *it is just a fact that we are responsible*. That's why the climate change issue is so central to this crisis of credibility. Unlike skepticism about, say, evolution or the efficacy of vaccinations or even moon landings, climate-change denial actually does involve a threat to "the world" per se, for which we are measurably responsible. And words like "crisis" and "threat" are appropriate because the way a genuine skepticism about science among masses of Americans "without college degrees"—the pollsters favorite rubric for Trump supporters—is converging with the cynical indifference of those who run fossil-fuel industries and buy politicians to support publicity campaigns dismissing the scientific consensus. That convergence is, as a matter of fact, leading to policies that may well be catastrophic—that will be if the whole goddam planet doesn't take action soon.

I'm suddenly reminded of an old friend—Paul Ryan, videographer and environmental activist—who loved to intone at appropriate moments, "The environment isn't an issue among others; it's the context for all the issues."

So this is undeniably an urgent matter, and all the more so because autocratic/populist identity politics seems to be ascending or ascendant all across the world. It isn't just the Trumpistas. And the tone and tenor of movements led by the likes of Marine Le Pen in France and Vladimir Putin in Russia, of course, but also Sebastian Kurtz in Austria, Recep Erdoğan in Turkey, Victor Orban in Hungary, Rodrigo Duterte in the Philippines, and Narendra Modi in India is not encouraging. If you want to see science play a significant role in describing the realities political leaders have to deal with, you should be very worried, because this is more than just short-sighted self-dealing; this is active opposition, indignant rebellion.

Ironically, though, the fact that climate change implicates "the world" so uniquely actually makes those who claim to know the truth, and who righteously assign responsibility on the basis of that knowledge, all the more inviting as targets for the skeptics. You can't look more elite than you look when you assert your personal responsibility for the whole world. I guess you know *everything*; who do you think you are? God?

In a CVS line a few months ago, I vaguely overheard a couple of amiable working class folks behind me, chatting about the weather. My ears perked up when I heard one of them, the woman, say—"That's why I don't believe all this global warming stuff. They can't even tell what the weather's going to be for the weekend." Her companion chortled appreciatively and said something about how they have to pretend to know because that's how they make a living. More chortling, knowing chortling—we've got your number.

I came within a hair of turning around and talking about sample sizes and statistics, but prudence prevailed. At the time, I thought the prudence was built into the etiquette of the situation. It would be plain rude to interrupt strangers in casual conversation and deliver a stern lecture of correction from a suddenly assumed podium of authority. On thinking back, though, I wonder if the prudence wasn't more deeply rooted. Suppose—wildly unlikely—but just suppose that I had managed not to be obnoxious and they had responded with something like open-minded interest? This was in a small town—suppose we ended up sitting in the diner, with me explaining the difference between weather forecasts and predictions of climate change? In that moment, looking back, I realized I wouldn't have done a very good job. I mean, I'm no expert.

Which leads to another complicating, yet essential, factor: people who really don't know *anything* about science, people who never read those astonishingly good *New Yorker* articles about this or that scientific enterprise—for them, scientists are lumped into the general "experts" category. And that's a real problem, because, as everybody knows, you can always find *some* credentialed "expert" who will say just about anything about

THE FREEDOM TO BELIEVE—OR NOT   263

anything. Especially "economists." They're scientists, right? I mean, all those formulas and graphs are right there, for all to see? And what about doctors? They're scientists too, right? And who doesn't have their favorite doctor-makes-hideous-blunder story? Probably one so dramatic that the fact that your blood pressure medication has been working for twenty-five years in spite of your poor lifestyle choices gets lost in that most ingenious of all hiding places—the utterly familiar. If you are a person "with a college degree," you likely know that there is huge difference between physics, or microbiology, or even geology—and economics. You don't need to know the actual science to know that; it comes with the degree, by a sort of osmosis.

So the problem isn't really with knowledge, expert or otherwise—the problem is with believing. And beliefs—in the age of Amazon, YouTube, Facebook, Twitter—are just resources that people use to fashion their identities. And there are so many beliefs to choose from! So you tend to go with the ones that make you feel good about yourself. And notice, by the way, that this principle applies to educated believers in science and not just to deniers. Think of the satisfaction you take in knowing that you aren't one of those idiots who doesn't "believe in" evolution but instead accepts an account of speciation based on scripture. It feels good to know that you are possessed of more elevated intentions informed by a better understanding of how things really are—where, once again, that doesn't have to mean that you know the actual science of chemistry, like a chemist, but just that you have a sense of the *kind* activity it is, of the *kind* of people and institutions that produce that kind of knowledge on an ongoing basis. It's a lot like the satisfaction you take in being astounded at the latest idiotic "alternative fact" being promoted in Trumpland.

As the reference to the Trumpian ethos suggests, this basic situation—this plethora of media-based belief options that defy coherence—has effects that go way beyond science deniers (or believers). Science denial can't be understood apart from that situation, however.

The attentive reader may have noticed that, in addition to expressing genuine distress with the willful ignorance on display among science deniers and their manipulators, this essay is suggesting that responsibility for this willful ignorance is deeply entrenched in our culture and that it implicates all of us in something that goes way beyond the scope of the rubric "without a college degree."

For starters, admit this, if the shoe fits: like me realizing I wouldn't be able to explain climate science very well if I had to, you very likely don't *really* understand the evidence behind the scientific consensus on climate change either, unless you happen to have made a special project out of it, for whatever reason. Unless you recently finished a *New Yorker* article summarizing that evidence the way only the *New Yorker* can—so that you haven't yet forgotten most of it—what do you really *know* about the science? Well, ok, let's see: "greenhouse gases," (I had forgotten until this moment what a clever little trope that is) are like a chemical ceiling that won't let, what?—carbon dioxide, I think, which makes air warmer for some reason?—won't let it escape the earth's atmosphere or, wait a minute, is it just that it won't let heat (which is actually nothing but particles of airy stuff, moving faster?) through the chemical ceiling? Oh, well, if you need to you can always Google it.

Then there's something about "core samples," I think, where they drill down into the tundra or somewhere with a hollow tube and then pull the tube out and, by looking at the layers in the tube or maybe "carbon dating" samples of the layers (you have no idea what that really is), or maybe examining certain traits of fossils in the layers—anyway, a lot of stuff like that—then they can tell what the climate was in these different time periods, going way back. And then, using statistical techniques you don't really understand either, they can tell how fast natural climate change happens and that tells you that now, thanks to those insidious greenhouse gases (air conditioners are bad; and spray deodorants, are they still bad?), things are heating up way faster than would be statistically likely otherwise.

What's really going on here, of course, is that you *trust* the science (as opposed to knowing it) and you trust (most of) the scientists so you choose to *believe* them. This is *not* to say that you aren't justified in trusting them, I think you are, I *know* you are—my point here is no way relativist. But the rather elaborate philosophical argument explaining

why (which I do know!) would take us too far afield; for now, just entertain this thought: thanks, as it were, to your college degree, you *choose* to
believe the science you don't really understand.

Which leads, finally and I think most importantly, to attitudes characteristic of the educated elite, attitudes that are—thanks to ubiquitous
media—now constantly on display, inviting linking and re-tweeting the
world over (Hilary's "basket of deplorables"). Those elites radiate feelings
of infinite superiority over those who have simply accepted the world
they were given, just like their parents and grandparents did. It's a point
that can't be made too often: these people may be ignorant, but they
aren't stupid. They can tell when they are being disdained. On the other
hand, we who are so habitually disdainful are more or less unaware of how
inflated our sense of our own importance actually is.

I teach a philosophy elective for seniors at a prestigious prep school and
I remember one young man, a brilliant fellow, especially gifted at mathematics and deeply invested in the hard sciences, physics above all. He
enjoyed the class but found a lot of it a bit too "soft" for his taste, a bit too
much like a history course or, God forbid, the poetry course he had been
required to take in tenth grade. But he liked an early essay by Nietzsche,
"On Truth and Lies in a Nonmoral Sense," especially the first few pages
where Nietzsche bears down ferociously on the place of Man in the great
scheme of things—the immensities of time and space that obliterate by
comparison that utterly inconsequential creature whose moment of existence is bound to pass in the blink of a cosmic eye. Nietzsche (his mouth
full of irony) cites Kant approvingly for his description of how humbling
it was to him personally, and mankind in general, to absorb the lessons of
the Copernican revolution.

This budding young scientist ate that up. And he had much to say on
how salutary that particular effect of science ought to be for the human
species, out of humility might come concern for the general well-being
and, yes, for the planet . . .

But I couldn't help but notice something (that Nietzsche missed,
could it be?) about the boy's manner as he was talking. So I interrupted

him in the middle of his description of how humbling it was for humans to be told that the earth was not at the center of universe and that they were a species of animal, like any other, and so on—I interrupted him and said, "You know what Russell, you don't sound very humble right now? I mean, you sound like you are riding pretty high, you sound like you are getting quite a kick out of being in a position to put mankind in his place—that is, down?"

Now, as I say, this was a fine young man—completely honest—and we had talked a lot about the sneaky ways Nietzsche's "will to power" finds to express itself, and his eyes glanced away for a few seconds as he looked within, for just as long as it took to realize that I was right; he lapsed immediately into a sheepish grin, but pleased as well—for he had learned something, and that would always have value for him.

So the expert, the scientist, and well-educated *New Yorker* readers have the courage to live with their "Copernican" understanding of man's negligible place in the universe and so, in all humility, forsake the illusory centralities and consolations of religion, which they leave behind for those "without college degrees" to cling to, along with their guns.

Really? Does it *still* take courage to live without God? Well, no doubt we note in passing now and then that we can make do without the "crutch" of religion. Does the thought of the immensities of time and space still provoke humility? Well, awe is a more likely response nowadays, I think—although even that is likely tinged with the merely "awesome," for such is the effect of seeing so many amazing movies and museum installations about the immensities of time and space. But humility? Not typically, surely; after all, it was us insignificant humans who developed the theories and technologies that have disclosed those immensities, that turned them into exhibits, as it were, exhibits of our own prowess as much as of the immensities themselves.

No doubt there is attitudinal variation here, person-to-person differences. But, if all that modesty is still an effect of science's revelations of our place in the great scheme of things, it is also a veneer that never quite succeeds in masking the enormous pride that attends the mastery of a

science and technology capable of penetrating nature's deepest secrets—
the pride of the God-killer.

I used to teach an anthropology course for seniors at that same prestigious prep school and, during the second semester, I organized students into research teams. We were trying to understand the origin of
moral relativism in the school. By the time the kids reached their senior
year some kind of relativism was the default position—but how did it
emerge, how did it take shape in earlier grades? How did it come to be
taken for granted that ethics comes down to personal opinion and social
upbringing and there's an end to it?

One group of researchers worked with fifth graders, a pivotal tran-
sitional group. They borrowed techniques from Lawrence Kohlberg,
who had taken Jean Piaget's basic ideas about cognitive development and
applied them to moral reasoning. One of the researchers used this scenario in his interviews: "You find a paper bag on the sidewalk and in it is
a lot of money. There are no hints as to its source, no sign of the original
owner. Just the bag and the money. What do you do—and how would
you justify your decision?"

Answers varied. Some said: put an ad in the paper; take it home to
parents; take it to the police; give it to charity; and, of course, there was
just keep it. One of the boys who said "just keep it" introduced his decision by saying that he knew others could make a different decision based
on their opinions of what was right, and that's what he was doing—and
who could say otherwise? The researching senior was inspired to say,
"What about God?"

To which this precocious eleven-year-old, after a moment's thought,
replied, "Well, I personally am an agnostic but, hypothetically, if there
was a God, I can see that he might have a different opinion from mine."

This story gets a big laugh when I tell it to groups of parents at the school.
They all recognize that sense of entitlement. Most students from these
families, in these schools, develop that sense. Adults have been attending
rapturously to their darling little pronouncements since they cooed their

first gurgles and have continue to attend, often with genuine fascination, to their expressions of opinion on anything and everything throughout their childhoods. It's part of the way people "with a college degree" express their love.

Of course, it's funny to hear an eleven-year-old place his own opinion on a par with God's opinion. We adults—having misjudged so many things for so long—know better. In fact, in that laughter, I detect a rueful note, a hint of "if only . . ." But, however that may be, science deniers (who are often religious) find in the demeanor of the expert, of the scientist, a reflection of that boy's assumption of the worth of his opinion. Except they don't think it's funny. They think it's arrogant and condescending.

But there is another dimension to this situation that also implicates people "with a college degree." It is an historical dimension, one that contributed fundamentally to the overall culture of optional belief that we all live in today, even as our anxiety about the dangers of science denial is making some of us realize that it may not always be a good thing for beliefs to be quite so optional. This particular feature of the situation demands special attention—and a special reckoning—from people who were there, back in the day, creating the counterculture at Woodstock, occupying the campus during the Cambodia bombings. I extend this anecdote a bit because it is, I believe, especially revealing. For those who remember, but also for younger people who didn't live through that era and might profit from a glimpse of the world that made their world:

The scene: a glorious summer day on Cape Cod, circa 1975. A dear friend of mine and her mother (lifelong bonds between us) are sitting on a screened-in porch, sipping lemonade. They have just returned from a three-day retreat at an ashram in the Berkshires and are telling me all about it. I was curious about meditation. I knew it was correlated with measurable physiological effects, and I had heard a lot about it from trustworthy people—I was thinking I might try it. So I was listening with genuine interest. But it soon became apparent that this particular retreat had ventured way beyond meditation. Apparently, the Maharishi (the one who hung out with the Beatles) had achieved a breakthrough. He and his closest disciples at the central ashram (in Geneva, was it?) had embarked upon a radical new phase of enlightenment. They had achieved something very special, soon to be revealed to the world at a mass meditation event that would channel peace vibes across the planet. But some of these highly secret new techniques had been shared with the fortunate participants at this particular retreat in the Berkshires, by way of preparation for the big day.

And what was this achievement? I asked.

Levitation.

I expressed a certain—ah—skepticism, shall we say? And they turned on me. These two extremely intelligent, Ivy League–educated women turned on me and proceeded to rake me over the coals with considerable vim. How could I be so arrogant and close-minded? How could I know that levitation wasn't possible?

As a general rule I make a practice of avoiding discussions of this kind; I smile and nod (no doubt a bit condescendingly) and wait for the subject to change. But this time I just snapped. I launched into a lecture on probability (I was reading a lot in philosophy of science at the time). I set up a scale to illustrate degrees of unlikelihood. At one extreme, logically impossible—a round square, given our definitions. Next, utterly, wildly unlikely, breaking established laws of physics at the Newtonian level—levitation. Next, very, very unlikely, given known laws—say, clairvoyance. And so on, through ESP and alien abduction and, toward the other end of the scale, the probability that Hitler and Eleanor Roosevelt were living together in Argentina.

All in vain. They believed.

Apparently they had been shown videotapes from the ashram in Geneva, and, though the tapes were rather grainy and shadowy for some reason, you could make out enough to tell that those yogis were...hovering.

Plus, my friends had practiced some of these techniques themselves in the Berkshires, don't forget, and, though they never actually got airborne, they could definitely feel themselves getting...lighter.

They wanted to believe. That's the key to this whole phenomenon, the thread to keep track of.

Exasperated, hoping to jolt them out of their post-retreat trance, I proposed a bet. Five hundred dollars each (a lot of money in those days, but I wasn't worried) that in two years levitation would be a universally acknowledged fact. Squadrons of levitating gurus floating down Fifth Avenue, on the Johnny Carson show, evidence like that.

And they took the bet!

Time passed, and no levitators materialized, but my dear friends were remarkably undisturbed. They weren't embarrassed into reassessing their worldview from the ground up, not at all, far from it. It was as if they had lost a bet on a sporting event. They had been wrong about the specifics on this one, but what did that prove? Maybe next time, who knows?

Who really knows anything?

Just recently I happened to tell one of my best graduate students that story. I was chortling as I went along, confident that she would share my point of view, but as I reached the end of my tale I noticed a little frown on her face—and she said, guess what?

"How do you know levitation isn't possible?"

It's this business of choice, you see, and it's not always a good thing. The irony of our time is that it is no longer "us," the radicals of yore, who are indulging in an orgy of choice. It is the populist forces of willful ignorance who carry the banner of transgression, who feel free to say anything, to believe anything. It is they who are challenging norms, violating categories, upending governance. We are the establishment, and science, perhaps more than any other institution, entitles us to whatever authority we have. But first, if we expect to get anywhere with that, we will have to learn how to talk to people who don't know as much as we do (which isn't that much). And we better figure out how to do that fast—there's a hard rain gonna fall.

LIST OF CONTRIBUTORS

Alba Alexander is Visiting Associate Professor in the Political Science Department in the University of Illinois at Chicago. She is coeditor of *Transforming the Local State* (Minnesota, forthcoming) and contributed to many journals, including *Harvard International Review*, *Congress and the Presidency*, and the *International Journal of Urban and Regional Research*.

Joseph Chuman has been the professional leader of the Ethical Culture Society of Bergen County, New Jersey, since 1974 and is a part-time leader of the Ethical Culture Society in New York City. He teaches courses in human rights in the Graduate School of Arts and Sciences at Columbia University and at Hunter College. He has also taught at the United Nations University for Peace in San Jose, Costa Rica; at Fairleigh Dickinson University; and at William Paterson University. In addition, Dr. Chuman is an activist, writer, and public speaker. He was the founder of the Northern New Jersey Sanctuary Coalition, which works with political asylum seekers, and has also worked on behalf of numerous other progressive causes. He has written for the *New York Times*, the *Record of Hackensack*, journals of opinion, academic texts, and several encyclopedias. He speaks to public audiences frequently on current issues, with an emphasis on the social, religious, and philosophical values they reflect.

Thomas de Zengotita teaches at the Draper Graduate Program at New York University and the Dalton School in New York City. He holds a BA, MA, and PhD in Anthropology from Columbia University. His interests include phenomenology, ethics, media theory, and modern intellectual history. He was contributing editor at *Harper's Magazine* from 2006 to 2011. Recent publications include "Ethics and Limits of Evolutionary Psychology" in the *Hedgehog Review*, Spring 2013; "On the Politics of Pastiche and Depthless Intensities: The Case of

Barack Obama" in the *Hedgehog Review*, Spring 2011. His next book, *Postmodern Theory and Progressive Politics* is due out from Palgrave Macmillan in 2019. He is now at work on a book for Stanford University Press called *A New Foundation for Human Rights: An Essay in Philosophical Anthropology*. His book *Mediated: How the Media Shapes Your World and the Way You Live in It* won the 2006 Marshall McLuhan Award.

Harrison Fluss is a lecturer of philosophy in New York City and corresponding editor for the journal *Historical Materialism*.

Barbara Forrest was, until her retirement in December 2017, a professor of philosophy in the Department of History and Political Science at Southeastern Louisiana University. She is the coauthor, with Paul R. Gross, of *Creationism's Trojan Horse: The Wedge of Intelligent Design* (Oxford University Press, 2007), and has authored and coauthored numerous articles in both academic and popular publications. She was an expert witness for the plaintiffs in the first legal case involving intelligent design, *Kitzmiller et al. v. Dover Area School District* (2005), which was resolved in favor of the plaintiffs. She previously served on the boards of directors of the National Center for Science Education and Americans United for Separation of Church and State.

Landon Frim is assistant professor of philosophy at Florida Gulf Coast University.

Margaret C. Jacob is Distinguished Professor of history at UCLA. She has a PhD from Cornell University and from the University of Utrecht. She is also a member of the American Philosophical Society and a fellow of the American Academy of Arts and Sciences. Her most recent book is *The First Knowledge Economy: Human Capital and the European Economy, 1750–1850* (Cambridge University Press, 2014).

Kurt Jacobsen is an associate (formerly research associate and lecturer) in the Program on International Politics, Economics and Security in the Political Science Department at the University of Chicago. He is the author or editor of eleven books, the producer of half a dozen documentaries, the book review editor at *Logos: A Journal of Modern Society and Culture*, and the coeditor of the *British Journal Free Associations: Psychoanalysis, Media, Groups and Politics*.

Diana M. Judd received her PhD from Rutgers University in New Brunswick, New Jersey. She is the author of many articles and book chapters, and her book *Questioning Authority: Political Resistance and the Ethic of Natural Science* has been recently re-released in paperback and digital form. Dr. Judd currently teaches political science at Metropolitan College of New York, in New York City.

Philip Kitcher was born in 1947 in London. He received his BA from Cambridge University and his PhD from Princeton. He has taught at several American universities and is currently John Dewey Professor of Philosophy at Columbia. He is the author of books on topics ranging from the philosophy of mathematics, the philosophy of biology, the growth of science, the role of science in society, naturalistic ethics, pragmatism, Wagner's *Ring*, Joyce's *Finnegans Wake*, and Mann's *Death in Venice*. He has been president of the American Philosophical Association (Pacific Division) and editor in chief of *Philosophy of Science*. A Fellow of the American Academy of Arts and Sciences, he was also the first recipient of the Prometheus Prize, awarded by the American Philosophical Association for work in expanding the frontiers of science and philosophy. He has been named a "Friend of Darwin" by the National Committee on Science Education, and received a Lannan Foundation Notable Book Award for *Living With Darwin*. He has been a Fellow at the Wissenschaftskolleg zu Berlin, where he was partially supported by a prize from the Humboldt Foundation, and in the autumn of 2015 he was the Daimler Fellow at the American Academy in Berlin. His most recent books are *Life After Faith: The Case for Secular Humanism* (Yale University Press, 2014), and *The Seasons Alter: How to Save our Planet in Six Acts*, coauthored with Evelyn Fox Keller (W. W. Norton, 2017). He is currently at work on a systematic version of Deweyan pragmatism, tentatively entitled *Progress, Truth, and Values*.

Michael Ruse is the Lucyle T. Werkmeister Professor of Philosophy and

Director of the History and Philosophy of Science Program at Florida State University. He is a Fellow of the Royal Society of Canada and a former Gifford Lecturer.

Lee Smolin is a theoretical physicist who works mainly on foundational problems such as quantum gravity and quantum theory. Born in New York City, he was educated at Hampshire College and Harvard University and served on the faculties of Yale, Syracuse, and Penn State Universities. Since 2001 he has been founding and senior faculty member at Perimeter Institute for Theoretical Physics. He has written or cowritten 199 scientific papers and is currently completing his fifth book exploring philosophical issues in physics and cosmology; he also coauthored a book on the reality of time with Roberto Mangabeira Unger. He lives with his family in Toronto.

Gregory Smulewicz-Zucker is a PhD student in the department of Political Science at Rutgers University and the managing editor of *Logos: A Journal of Modern Society and Culture* (www.logosjournal.com). He holds degrees in history, philosophy, and political science from the University of Cambridge, the Graduate Center, CUNY, and Rutgers University. His most recent edited books are *The Political Thought of African Independence: An Anthology of Sources* (Hackett Publishing) and, with Michael J. Thompson, *Radical Intellectuals and the Subversion of Progressive Politics* (Palgrave).

Alan Sokal is professor of mathematics at University College, London, and professor of physics at New York University.

Michael J. Thompson is professor of political theory in the Department of Political Science at William Paterson University. His forthcoming books include *The Specter of Babel: Toward a Reconstruction of Political Judgment* (SUNY Press) and *Twilight of the Self: The Eclipse of Autonomy in Modern Society* (Stanford University Press). He is also the founding editor of *Logos: A Journal of Modern Society and Culture*.

NOTES

Chapter 1: What Is Science and Why Should We Care?

*This essay was originally published in *Logos* 12, no. 2 (Spring 2013) and is used here with permission. The essay is an expanded version of a talk given at the third Sense About Science Annual Lecture, University College, London, February 27, 2008.

1. It is crucial, in order to avoid misunderstandings, that the word "scientific" here be understood in the broad sense to be explained below, namely as "investigations aimed at acquiring accurate knowledge of factual matters relating to any aspect of the world by using rational empirical methods analogous to those employed in the natural sciences." Alternatively, one could use the phrase "evidence-based worldview."

2. George Orwell, "Politics and the English Language," [1946] in *A Collection of Essays* (New York: Harcourt Brace Jovanovich, 1953), pp. 156–71.

3. Jean Bricmont, "Préface," in Alan Sokal, *Pseudosciences et postmodernisme: Adversaires ou compagnons de route?* (Paris: Odile Jacob, 2005), pp. 7–38.

4. Susan Haack, *Evidence and Inquiry: Towards Reconstruction in Epistemology* (Oxford, UK: Blackwell, 1993); Susan Haack, *Defending Science—Within Reason: Between Scientism and Cynicism* (Amherst, NY: Prometheus Books, 2003); Susan Haack; *Manifesto of a Passionate Moderate: Unfashionable Essays* (Chicago: University of Chicago Press, 1998).

5. If you, by contrast, prefer to restrict the term "science" to the natural sciences only, then it suffices to replace the word "science" everywhere in my text by the phrase "investigations aimed at acquiring accurate knowledge of factual matters relating to any aspect of the world by using rational empirical methods analogous to those employed in the natural sciences."

6. Haack, *Evidence and Inquiry*; Haack, *Defending Science*; Haack, *Manifesto of a Passionate Moderate*.

7. Kenneth J. Gergen, "Feminist Critique of Science and the Challenge of Social Epistemology," in *Feminist Thought and the Structure of Knowledge*, ed. Mary McCanney Gergen (New York: New York University Press, 1988), pp. 27–48.

8. Harry M. Collins, "Stages in the Empirical Programme of Relativism," *Social Studies of Science* 11, no. 1 (February 1981): 3; Jean Bricmont and Alan Sokal, "Science and Sociology of Science: Beyond War and Peace," in *The One Culture? A Conversation about Science*, ed. Jay Labinger and Harry Collins (Chicago: University of Chicago Press, 2001), pp. 27–47, 179–83, and 243–54 [also available online at http://www.physics.nyu.edu/faculty/sokal/collins_v4b _clean.pdf]; Jean Bricmont and Alan Sokal, "Reply to Gabriel Stolzenberg," *Social Studies of Science* 34, no. 1 (February 2004): 107–13 [also available online at http://www.physics.nyu .edu/faculty/sokal/reply_to_stolzenberg_v2.pdf].

9. Barry Barnes and David Bloor, "Relativism, Rationalism, and the Sociology of Knowledge," in *Rationality and Relativism*, ed. Martin Hollis and Steven Lukes (Oxford: Blackwell, 1982), pp. 21–47.

10. Bruno Latour, *Science in Action: How to Follow Scientists and Engineers through Society* (Cambridge, MA: Harvard University Press, 1987), p. 99.

11. Stanley Aronowitz, *Science as Power: Discourse and Ideology in Modern Society* (Minneapolis: University of Minnesota Press, 1988), p. 204.

12. N. Katherine Hayles, "Gender Encoding in Fluid Mechanics: Masculine Channels and Feminine Flows," *Differences: A Journal of Feminist Cultural Studies* 4, no. 2 (1992): 16–44.

13. Andrew Pickering, *Constructing Quarks: A Sociological History of Particle Physics* (Chicago: University of Chicago Press, 1984), p. 413; Eugene P. Wigner, "The Unreasonable Effectiveness of Mathematics in the Natural Sciences," *Communications on Pure and Applied Mathematics* 13, no. 1 (February 1960): 1–14.

14. James Robert Brown, *Who Rules in Science? An Opinionated Guide to the Wars* (Cambridge, MA: Harvard University Press, 2001).

15. And don't even get me started on Trump and his friends.

16. Bruno Latour and Steve Woolgar, *Laboratory Life: The Social Construction of Scientific Facts* (Beverly Hills: Sage Publications, 1979).

17. Chris Mooney, *The Republican War on Science* (New York: Basic Books, 2005).

18. Bruno Latour, "Why Has Critique Run Out of Steam? From Matters of Fact to Matters of Concern," *Critical Inquiry* 30, no. 2 (Winter 2004): 225–48.

19. Noam Chomsky, "Rationality/Science," *Z Papers Special Issue on Postmodernism and Rationality*, 1992, https://zcomm.org/wp-content/uploads/ScienceWars/sciencechomreply .htm; Michael Albert, "Not All Stories Are Equal," *Z Papers Special Issue on Postmodernism and Rationality*, 1992, https://zcomm.org/wp-content/uploads/ScienceWars/notallstories. htm; Noam Chomsky, *Year 501: The Conquest Continues* (Boston: South End Press, 1993); Barbara Ehrenreich, "For the Rationality Debate," *Z Papers Special Issue on Post-Modernism and Rationality*, 1992, https://zcomm.org/wp-content/uploads/ScienceWars/ehrenscience.htm.

20. Alan Sokal, "Pseudoscience and Postmodernism: Antagonists or Fellow-Travelers?" in *Archaeological Fantasies: How Pseudoarchaeology Misrepresents the Past and Misleads the Public*, ed. Garrett G. Fagan (Abingdon, UK: Routledge, 2006), pp. 286–361 [also available online at http://www.physics.nyu.edu/faculty/sokal/pseudoscience_rev.pdf]; Alan Sokal, *Beyond the Hoax: Science, Philosophy, and Culture* (New York: Oxford University Press, 2008).

21. Jos Kleijnen, Paul Knipschild, and Gerben ter Riet, "Clinical Trials of Homoeopathy," *British Medical Journal* 302 (February 9, 1991): 316–23, and "Correction: Clinical Trials of Homoeopathy," *British Medical Journal* 302 (April 6, 1991): 818; Klaus Linde, Nicola Clausius, Gilbert Ramirez, et al., "Are the Clinical Effects of Homoeopathy Placebo Effects? A Meta-Analysis of Placebo-Controlled Trials," *Lancet* 350, no. 9081 (September 20, 1997): 834–43; Klaus Linde, Michael Scholz, Gilbert Ramirez, et al., "Impact of Study Quality on Outcome in Placebo-Controlled Trials of Homeopathy," *Journal of Clinical Epidemiology* 52, no. 7 (July 1999): 631–36; Klaus Linde and Dieter Melchart, "Randomized Controlled Trials of Individualized Homeopathy: A State-of-the-Art Review," *Journal of Alternative and Complementary Medicine* 4, no. 4 (December 1998): 371–88; Michel Cucherat, Margaret C. Haugh, Mary Gooch, et al., "Evidence of Clinical Efficacy of Homeopathy: A Meta-Analysis of Clinical Trials," *European Journal of Clinical Pharmacology* 56, no. 1 (April 2000): 27–33; Aijing Shang, Karin Huwiler-Müntener, Linda Nartey, et al., "Are the Clinical Effects of Homeopathy Placebo Effects? Comparative Study of Placebo-Controlled Trials of Homoeopathy and Allopathy," *Lancet* 366, no. 9487 (August 27, 2005): 726–32; Edzard Ernst, "A Systematic Review of Systematic Reviews of Homeopathy," *British Journal of Clinical Pharmacology* 54, no. 6 (December 2002): 577–82; Kenneth F. Schulz, Iain Chalmers, Richard J. Hayes, et al., "Empirical Evidence of Bias: Dimensions of Methodological Quality Associated with Estimates of Treatment Effects in Controlled Trials," *Journal of the American Medical Association* 273, no. 5 (February 1, 1995): 408–12; Khalid S. Khan, Salim Daya, and Alejandro R. Jadad, "The Importance of Quality of Primary Studies in Producing Unbiased Systematic Reviews," *Archives of Internal Medicine* 156, no. 6 (March 25, 1996): 661–66; David Moher, Ba' Pham, Alison Jones, et al., "Does Quality of Reports of Randomised Trials Affect Estimates of Intervention Efficacy Reported in Meta-Analyses?" *Lancet* 352, no. 9128 (August 22, 1998): 609–13; Rudolf W. Poolman,

Peter A. A. Struijs, Rover Krips, et al., "Reporting of Outcomes in Orthopaedic Randomized Trials: Does Blinding of Outcome Assessors Matter?" *Journal of Bone and Joint Surgery* 89, no. 3 (March 2007): 550–58; Kenneth F. Schulz, "Assessing Allocation Concealment and Blinding in Randomised Controlled Trials: Why Bother?" *Evidence-Based Medicine* 5, no. 2 (March 2000): 36–38.

22. Linde, Clausius, Ramirez, et al., "Are the Clinical Effects of Homoeopathy Placebo Effects?"; Linde, Scholz, Ramirez, et al., "Impact of Study Quality"; Paul Seed, "Correspondence: Meta-Analysis of Homoeopathy Trials," *Lancet* 351, no. 9099 (January 31, 1998): 365; Klaus Linde and Wayne B. Jonas, "Meta-Analysis of Homoeopathy Trials: Authors' Reply," *Lancet* 351, no. 9099 (January 31, 1998): 367–68; Ernst, "Systematic Review of Systematic Reviews"; Shang, Huwiler-Müntener, Nartey, et al., "Are the Clinical Effects of Homoeopathy Placebo Effects?" (see also letters to the editor in the *Lancet* 366, no. 9503 (December 17, 2005): 2081–86, along with a reply from the authors that gives information that was unfortunately omitted from the original report).

23. Select Committee on Science and Technology, "Minutes of Evidence: Examination of Witnesses (Questions 520–39)" (testimony of Jonathan Brostoff, Kate Chatfield, Chris Corrigan, and Edzard Ernst), UK House of Lords, February 21, 2007, http://www.publications.parliament.uk/pa/ld200607/ldselect/ldsctech/166/7022105.htm.

24. US Food and Drug Administration, *Compliance Policy Guide Section 400.400: Conditions under Which Homeopathic Drugs May Be Marketed* (Silver Spring, MD: US Department of Health and Human Services, 2010), http://www.fda.gov/ICECI/ComplianceManuals/CompliancePolicyGuidanceManual/ucm074360.htm.

25. "Explanatory Memorandum to the Medicines for Human Use (National Rules for Homoeopathic Products) Regulations 2006," UK Medicines and Healthcare Products Regulatory Agency, July 19, 2006, http://www.legislation.gov.uk/uksi/2006/1952/memorandum/contents; "The Medicines for Human Use (National Rules for Homoeopathic Products) Regulations 2006," UK Secretary of State for Health, Statutory Instrument 2006, no. 1952, July 19, 2006, came into force September 1, 2006, http://www.legislation.gov.uk/uksi/2006/1952/contents/made.

26. Theodosius Dobzhansky, "Nothing in Biology Makes Sense except in the Light of Evolution," *American Biology Teacher* 35, no. 3 (March 1973): 125–29.

27. Pardon the pun.

28. Anita Miller, ed., *George W. Bush versus the US Constitution* (Chicago: Academy Chicago, 2006).

29. Pius XII, *Humani Generis: Encyclical of Pope Pius XII Concerning Some False Opinions Threatening to Undermine the Foundations of Catholic Doctrine*, August 12, 1950, http://w2.vatican.va/content/pius-xii/en/encyclicals/documents/hf_p-xii_enc_12081950_humani-generis.html; Salman Hameed, "Bracing for Islamic Creationism," *Science* 322, no. 5908 (December 12, 2008): 1637–38; Riaz Hassan, "On Being Religious: Patterns of Religious Commitment in Muslim Societies," *Muslim World* 97, no. 3 (July 2007): 437–78.

30. Stephen Jay Gould, *Rocks of Ages: Science and Religion in the Fullness of Life* (New York: Ballantine, 1999).

31. Sokal, "Pseudoscience and Postmodernism: Antagonists or Fellow-Travelers," chapter 8 in *Beyond the Hoax*, pp. 263–370.

32. Jean Bricmont, "Science et religion: l'irréductible antagonism," in *Ou va Dieu?* ed. Antoine Pickels and Jacques Sojcher (Brussels: Revue de l'Université de Bruxelles, Editions Complexe, 1999), pp. 247–64; reprinted in *Agone* 23 (2000): 131–51, and in Jean Dubessy and Guillaume Lecointre, eds., *Intrusions spiritualistes et impostures intellectuelles en sciences* (Paris: Editions Syllepse, 2001), pp. 121–38.; also available online at http://www.dogma.lu/txt/JB-Science01.htm.

33. Sam Harris, *The End of Faith: Religion, Terror, and the Future of Reason* (New York: W. W. Norton, 2004).

34. John Paul II, *Encyclical Letter Fides et Ratio of the Supreme Pontiff John Paul II to the Bishops of the Catholic Church on the Relationship between Faith and Reason* (Washington, DC: United States Catholic Conference, September 14, 1998), http://w2.vatican.va/content/john -paul-ii/en/encyclicals/documents/hf_jp-ii_enc_14091998_fides-et-ratio.html.

35. Sokal, "Pseudoscience and Postmodernism," in *Beyond the Hoax.*

36. Norman Levitt, "Response to Freudenberg," *Technoscience: Newsletter of the Society for Social Studies of Science* 9, no. 2 (Spring 1996).

37. John Prados, *Hoodwinked: The Documents That Reveal How Bush Sold Us a War* (New York: New Press, 2004); Miller, ed., *George W. Bush versus the US Constitution*; Frank Rich, *The Greatest Story Ever Sold: The Decline and Fall of Truth in Bush's America* (New York: Penguin, 2007); Michael Smith, "Blair Planned Iraq War from Start," *Sunday Times* (London), May 1, 2005, [The complete texts of the publicly available Downing Street Memos that Smith discusses in the *Sunday Times* article can be found at http://downingstreetmemo.com/]; US House of Representatives Committee on Government Reform, "Iraq on the Record: The Bush Administration's Public Statements on Iraq," prepared for Representative Henry A. Waxman by the Special Investigations Division of the Minority Staff of the Committee on Government Reform (Washington, DC: Minority Staff Special Investigations Division, 2004), https://www .hsdl.org?abstract&did=445160.

38. *The Compact Edition of the Oxford English Dictionary, Volume I: A-O* (Oxford: Oxford University Press, 1971), p. 1073. See also Henry Campbell Black, *A Treatise on the Rescission of Contracts and Cancellation of Written Instruments*, vol. 1 (Kansas City, MO: Vernon Law, 1916); Henry Campbell Black, *Black's Law Dictionary: Definitions of the Terms and Phrases of American and English Jurisprudence, Ancient and Modern*, ed. Joseph R. Nolan and Michael J. Connolly, 5th ed. (St. Paul, MN: West Publishing, 1979); Denis Lane McDonnell and John George Monroe, *Kerr on the Law of Fraud and Mistake*, 7th ed. (London: Sweet & Maxwell, 1952).

39. See George Spencer Bower and K. R. Handley, *Actionable Misrepresentation*, 4th ed. (London: Butterworths, 2000).

40. R. F. V. Heuston and R. A. Buckley, *Salmond and Heuston on the Law of Torts*, 21st ed. (London: Sweet & Maxwell, 1996); Iraq Family Health Survey Study Group [Amir H. Alkhuzai et al.], "Violence-Related Mortality in Iraq from 2002 to 2006," *New England Journal of Medicine* 358, no. 5 (January 31, 2008): 484–93.

41. "[For there to be fraud] there must be a misstatement of an existing fact: but the state of a man's mind is as much a fact as the state of his digestion. It is true that it is very difficult to prove what the state of a man's mind at a particular time is, but if it can be ascertained it is as much a fact as anything else. A misrepresentation as to the state of a man's mind is, therefore, a misstatement of fact" [Edgington v. Fitzmaurice (1885) 29 Ch D 459 at 483, CA, per Bowen LJ].

42. "If the facts are not equally known to both sides, then a statement of opinion by the one who knows the facts best involves very often a statement of a material fact, for he impliedly states that he knows facts which justify his opinion" [Smith v. Land and House Property Corporation (1885) 28 Ch D 7 at 15, per Bowen LJ]; American Law Institute, *Restatement of the Law, Second, Torts 2d*, vol. 3 (St. Paul, MN: American Law Institute, 1986); William Lloyd Prosser, *Handbook of the Law of Torts*, 4th ed. (St. Paul, MN: West Publishing, 1971).

43. Gilbert Burnham, Riyadh Lafta, Shannon Doocy, and Les Roberts "Mortality after the 2003 Invasion of Iraq: A Cross-Sectional Cluster Sample Survey," *Lancet* 368, no. 9545 (October 21, 2006): 1421–28. See also letters to the editor and authors' reply, in the *Lancet* 369, no. 9556 (January 13, 2007): 101–105.

44. Amy Belasco, *The Cost of Iraq, Afghanistan, and Other Global War on Terror Operations Since 9/11* (Washington, DC: Congressional Research Service, December 8, 2014), http://www .fas.org/sgp/crs/natsec/RL33110.pdf; Peter Orszag, *Estimated Costs of US Operations in Iraq and Afghanistan and of Other Activities Related to the War on Terrorism* (testimony before the Committee on the Budget, US House of Representatives; Washington, DC: Congressional Budget

Office, October 24, 2007), http://www.cbo.gov/publication/19202; Scott Wallsten and Katrina Kosec, "The Economic Costs of the War in Iraq" (working paper 05-19, AEI-Brookings Joint Center for Regulatory Studies, Washington, DC, September 2005), https://papers.ssrn.com/sol3/papers.cfm?abstract_id=848408; Linda Bilmes and Joseph E. Stiglitz, *The Three Trillion Dollar War: The True Cost of the Iraq Conflict* (New York: W. W. Norton, 2008); David Leonhardt, "What $1.2 Trillion Can Buy," *New York Times*, January 17, 2007.

45. American Law Institute, *Restatement of the Law, Second, Torts 2d*, vol. 3, as adopted and promulgated by the American Law Institute in Washington, DC, May 19, 1976 (St. Paul, MN: American Law Institute, 1977); Prosser, *Handbook of the Law of Torts*.

46. Prados, *Hoodwinked*; Miller, ed., *George W. Bush*; Rich, *Greatest Story Ever Sold*; Smith, "Blair Planned Iraq War"; US House of Representatives, "Iraq on the Record."

47. One of the additional corrupting effects of cynicism is that it undermines our ability to properly appreciate those politicians who do have the courage to tell us the truth—even when it is unsettling, and even when it contradicts our (and their own) preconceptions.

48. Haack, *Manifesto*.

49. Many postmodernists reject the fact-value distinction, but I strongly uphold it.

50. Bricmont, "Préface," in Sokal, *Pseudosciences et postmodernisme*.

Chapter 2: Science and the Democratic Mind

1. Stanley Aronowitz, *Science as Power: Discourse and Ideology in Modern Society* (Minneapolis: University of Minnesota Press, 1988), p. 6.

2. John Kurt Jacobsen, *Technical Fouls: Democratic Dilemmas and Technological Change* (Boulder, CO: Westview, 2000), p. 5.

3. See the important study by Diana Judd, *Questioning Authority: Political Resistance and the Ethic of Natural Science* (New Brunswick, NJ: Transaction Publishers, 2009).

4. Thomas Paine, *The Rights of Man* (London: W. T. Sherwin, 1817), p. 114.

5. Thomas Jefferson, *Notes on the State of Virginia* (Richmond, VA: J. W. Randolph, 1853), p. 178.

6. Carl Sagan, *The Demon-Haunted World: Science as a Candle in the Dark* (New York: Ballantine Books, 1996), p. 416.

7. Gunnar Myrdal, *Objectivity in Social Research* (New York: Pantheon Books, 1969), p. 29.

8. Immanuel Kant, *Critique of Practical Reason* (New York: Cambridge University Press, 1996), p. 135.

9. Aristotle, *Politics*, 1277^{b}III: 10–15, Oxford Classical Texts (Oxford: Oxford University Press., 1932).

10. Milton Rokeach, *The Open and Closed Mind: Investigations into the Nature of Belief Systems and Personality Systems* (New York: Basic Books, 1960), p. 40. Also see the more philosophical discussion by David Weissman, *Truth's Debt to Value* (New Haven, CT: Yale University Press, 1993), pp. 101ff.

11. Plato, *Republic*, 533^d, Oxford Classical Texts (Oxford: Oxford University Press., 1934).

12. Jean-Jacques Rousseau, *Emile: or On Education*, trans. Allan Bloom (New York: Basic Books, 1979), p. 272.

13. Paul Feyerabend, *Science in a Free Society* (London: Verso, 1978), p. 9.

14. See Max Weber, "Science as a Vocation," in *From Max Weber: Essays in Sociology*, ed. Hans Gerth and C. Wright Mills (New York: Oxford University Press, 1946), pp. 129–56.

15. Leo Tolstoy, *A Confession and Other Religious Writings* (London: Penguin, 1987), p. 132.

16. Stanley Aronowitz, "Postmodernism and Politics," in *Universal Abandon? The Politics of Postmodernism*, ed. Andrew Ross (Minneapolis: University of Minnesota Press, 1988), pp. 50–51.

17. See the discussion by Jacques Ellul, *The Technological Society*, trans. John Wilkinson (New York: Vintage Books, 1964), pp. 79ff.

18. Neil Postman, *Technopoly: The Surrender of Culture to Technology* (New York: Vintage, 1992), p. 55.

Chapter 3: The Synthesis of Science and Democracy: A Deweyan Appraisal

1. The evolution of illiberal trends in various dimensions of American society is subject to a burgeoning literature in recent years. Representative is *Why the Rights Went Wrong: Conservatism—From Goldwater to Trump and Beyond*, by journalist E. J. Dione (New York: Simon and Schuster, 2016), which chronicles the spiraling extremism in the Republican Party since the candidacy of Barry Goldwater. *Fantasyland: How America Went Haywire: A 500-Year History*, by Kurt Andersen (New York: Random House, 2017), is an extensive compendium documenting the evolution of the slide into irrationalism evident in America since its founding.

2. Steven C. Rockefeller, *John Dewey: Religious Faith and Democratic Humanism* (New York: Columbia University Press, 1991), p. 400.

3. Joseph Chuman, *Speaking of Ethics: Living a Humanist Life* (North Charleston, SC: CreateSpace, 2014), p. 219.

4. Ibid., p. 222.

5. This view, consistent with his naturalism, situates Dewey firmly among the empiricists and in opposition to the idealist philosophic tradition as well as to absolutism. Clearly, Darwin was a major influence in this regard. "For Dewey, moral relativism was a direct consequence of Darwinian evolutionary theory. As he interpreted it, Darwinian thought rejects all ideas of first and final causes and emphasizes the pervasive presence of change." (Rockefeller, *John Dewey*, p. 284.)

6. Rockefeller, *John Dewey*, p. 284.

7. John Dewey, *The Quest for Certainty: A Study of the Relation of Knowledge and Action* (New York: G. P. Putnam's Sons, 1929), pp. 204–205.

8. Ibid., p. 213.

9. Joseph Ratner, ed., *Intelligence in the Modern World: John Dewey's Philosophy* (New York: Modern Library, 1939), p. 632.

10. Robert Westbrook, *John Dewey and American Democracy* (Ithaca and London: Cornell University Press, 1991), p. 141.

11. John Dewey, *Freedom and Culture* (New York: G. P. Putnam's Sons, 1939), p. 145.

12. Ibid., p. 146.

13. Ibid., p. 148.

14. James Gouinlock, *John Dewey's Philosophy of Value* (New York: Humanities Press 1972), p. 348.

15. Rockefeller, *John Dewey*, p. 4. Dewey reconstructs religion within a non-supernatural framework by emphasizing "the religious" as opposed to "religion." See John Dewey, *A Common Faith* (New Haven, CT: Yale University Press, 1971).

16. Richard Rorty, *Achieving Our Country: Leftist Thought in Twentieth-Century America* (Cambridge: Harvard University Press, 1998), p. 18.

17. Ibid.

18. Dewey's philosophy of the individual bears a striking parallel to his part-time colleague at Columbia Felix Adler, who was also the founder of the Ethical Culture movement. Metaphysically, Adler was a neo-Kantian idealist, while Dewey was the foremost proponent of naturalism. But their social and political philosophies were very similar. Adler envisioned individuals as members of an infinite spiritual organism, which contained all other members joined in mutually reciprocal relations. Remarking on the nature of individualism, Adler

once stated that individuals are "monads with windows," and thereby affirmed the self as both individual and social simultaneously.

19. Westbrook, *John Dewey*, pp. 445–47.

20. Alan Ryan, *John Dewey and the High Tide of American Liberalism* (New York: W. W. Norton, 1995), p. 19.

21. Rockefeller, *John Dewey*, p. 240.

22. Sarah Jaffee, *Necessary Trouble: Americans in Revolt* (New York: Nation Books, 2016).

23. Westbrook, *John Dewey*, p. 552.

Chapter 4: The Philosophy of the Open Future

1. This essay has been in progress since 2002. Parts of it have served as first drafts of talks at TED, ideacity, and elsewhere, as well as in the epilogue of *Time Reborn*, but the whole, showing the relationship between the different concerns, has not so far been published.

2. This economic notion of equilibrium has more in common with an equilibrium of forces in statics than it does with thermodynamic equilibrium.

3. Pia Nandini Malaney and Eric Weinstein, "Welfare Implications of Divisia Indices," chap. 2 in "The Index Number Problem: A Differential Geometric Approach," by Pia Malaney (PhD diss., Harvard University, 1996), http://leesmolin.com/wp-content/uploads/2013/06/MalaneyThesis.pdf.

See also Eric Weinstein, "Neo-Classical Economics and Gauge Theory," Eric-Weinstein.net, http://www.eric-weinstein.net/economictheory.html; Eric Weinstein, "Gauge Theory and Inflation: Enlarging the Wu-Yang Dictionary to a Unifying Rosetta Stone for Geometry in Application" (seminar, Perimeter Institute, May 24, 2006), https://www.youtube.com/watch?v=h5gnATQMtPg; and Eric Weinstein, "The Practical Side of 'Pure Mathematics': How Differential Geometry Can Save the US Government 40 Billion Dollars a Year," abstract, September 25, 1997, http://www.math.dartmouth.edu/~colloq/f97/weinstein.html.

4. Samuel E. Vázquez, "Scale Invariance, Bounded Rationality, and Non-Equilibrium Economics" (paper, Perimeter Institute for Theoretical Physics, February 23, 2009), arXiv:0902.3840v1, available at http://arxiv.org/pdf/0902.3840.pdf.

5. Charles Sanders Peirce, *Collected Papers*, vol. 6, *Scientific Metaphysics*, ed. Charles Hartshorne and Paul Weiss (Cambridge, MA: Belknap Press, 1934), p. 15.

6. See Robert Mangabeira Unger, *The Self Awakened: Pragmatism Unbound* (Cambridge, MA: Harvard University Press, 2007).

7. Evidence that anyone may see or examine, i.e., not private revelations.

8. Stewart Brand, *Whole Earth Discipline: An Ecopragmatist Manifesto* (New York: Atlantic Books, 2010).

Chapter 5: The Scientific Revolution and Individual Inquiry

1. Aristotle, *The Politics*, trans. Carnes Lord (Chicago: University of Chicago Press, 1984), p. 118.

2. Ibid., p. 126.

3. Alexander Hamilton, James Madison, and John Jay, *The Federalist Papers*, no. 10 (New York: Penguin, 1961), p. 76.

4. Ibid.

5. Ibid., p. 75.

6. Ibid., no. 51, p. 319.

7. Ibid.

8. Francis Bacon, *The Advancement of Learning* (1605: Whitefish, MT: Kessinger Publishing, 1994), p. 12.

9. Ibid., p. 17.

10. Francis Bacon, *The New Organon*, Aphorism 89 (Indianapolis: Bobbs-Merrill, 1960), p. 87.

11. Ibid., p. 43.

12. Francis Bacon, *Essays, or Councils, Civic and Moral* (Hertfordshire, UK: Wordsworth Editions, 1997), pp. 5–6.

13. Scholars agree that this piece was probably written between the years 1622–33. It was published posthumously by Bacon's chaplain in 1629.

14. Bacon, *New Organon*, p. 39.

15. Ibid., p. 29.

16. Ibid., p. 13.

17. Ibid., pp. 20–21.

18. Ibid.

19. Ibid., pp. 48–49.

20. Ibid.

21. Bacon, *Essays*, p. 6.

22. Bacon, *New Organon*, p. 52.

23. Ibid., p. 88.

24. Ibid., p. 74.

25. John Locke, "Second Treatise of Government," in *Two Treatises of Government* (Cambridge: Cambridge University Press, 1998), paragraph 3.

26. Ibid., paragraph 6.

27. Ibid., paragraph 61.

28. Ibid., paragraph 4.

29. Ibid., paragraph 22.

30. Ibid.

31. Ibid., paragraph 57 (emphasis in the original).

32. Ibid.

33. Ibid., paragraph 132.

34. Ibid., paragraph 134.

35. Ibid., paragraph 99.

36. Ibid., and for more, see Locke's discussion in Aphorism XIX.

37. Ibid., paragraph 220.

38. Ibid., paragraph 240.

39. Ibid., paragraph 225.

40. Ibid., paragraph 199.

Chapter 6: The Left, Science Studies, and Global Warming

1. Sir Isaac Newton, *De Gravitatione et Aequipondio Fluidorum*, trans. W. B. Allen, http://williambarclayallen.com/translations/De_Gravitatione_et_Aequipondio_Fluidorum_translation.pdf: "If we say with Descartes (that) extension is body, do we not rather manifestly spread the way to atheism, for then that extension is not being created but was from eternity, whereupon we have an absolute idea of it without any relation to God, and thus we are able to conceive existence for the time being as if at that time we would suppose God not to be. And no distinction of mind from body, according to this philosophy, is understandable, lest simultaneously we say that mind is by no means extension, and thus is substantially present in no extension, or is no place; and so too if we say it is not that by means of which it is seen; however, I have plainly

restored its minimum understandable union with body, not saying (it is) impossible. Moreover, if the distinction of substances into thinking and extended is lawful and perfect, then God does not eminently contain, and hence he cannot create, extension within himself; but God and extension are two substances severally called absolutely complete and singular."

2. Bruno Latour, *We Have Never Been Modern*, trans. Catherine Porter (Cambridge, MA: Harvard University Press, 1993).

3. Bruno Latour, "One More Turn after the Social Turn: Easing Science Studies into the Non-Modern World," in *The Social Dimensions of Science*, ed. Ernan McMullin (South Bend, IN: University of Notre Dame Press, 1992), p. 289. These points are discussed in greater detail in Margaret C. Jacob, "Reflections on Bruno Latour's Version of the Seventeenth Century," in *A House Built on Sand: Exposing Postmodernist Myths about Science*, ed. Noretta Koertge (New York: Oxford University Press, 1998), pp. 240–54.

4. Bruno Latour, "Why Has Critique Run Out of Steam? From Matters of Fact to Matters of Concern," *Critical Inquiry* 30, no. 2 (Winter 2004): 227: "Entire PhD programs are still running to make sure that good American kids are learning the hard way that facts are made up, that there is no such thing as natural, unmediated, unbiased access to truth, that we are always prisoners of language, that we always speak from a particular standpoint, and so on, while dangerous extremists are using the very same argument of social construction to destroy hard-won evidence that could save our lives. Was I wrong to participate in the invention of this field known as science studies? Is it enough to say that we did not really mean what we said? Why does it burn my tongue to say that global warming is a fact whether you like it or not? Why can't I simply say that the argument is closed for good? Should I reassure myself by simply saying that bad guys can use any weapon at hand, naturalized facts when it suits them and social construction when it suits them? Should we apologize for having been wrong all along? Or should we rather bring the sword of criticism to criticism itself and do a bit of soul-searching here: what were we really after when we were so intent on showing the social construction of scientific facts?"

For the history of science studies at one of its founding centers see, "Science Studies Unit," Science, Technology, and Innovation Studies, School of Social and Political Science, University of Edinburgh, 2017, http://www.stis.ed.ac.uk/about/history/science_studies_unit.

5. Latour, "Why Has Critique Run Out of Steam?", pp. 231–32.

6. Bruno Latour, "Two Bubbles of Unrealism: Learning from the Tragedy of Trump," *Los Angeles Review of Books*, November 17, 2016.

7. Ibid.

8. Steve Paulson, "The Critical Zone of Science and Politics: An Interview with Bruno Latour," *Los Angeles Review of Books*, February 23, 2018.

9. See Graham Harman, *The Prince of Networks: Bruno Latour and Metaphysics* (Melbourne, Australia: Re.Press, 2009), pp. 58–61. For a different perspective, see Jacob, "Reflections on Bruno Latour's Version of the Seventeenth Century"; Paulson, "Critical Zone of Science and Politics."

10. Spencer Weart, *The Discovery of Global Warming*, February 2018, https://history.aip.org/climate/index.htm.

11. Peter Baker and Peter Slevin, "Bush Remarks on 'Intelligent Design' Theory Fuel Debate," *Washington Post*, August 3, 2005:

President Bush invigorated proponents of teaching alternatives to evolution in public schools with remarks saying that schoolchildren should be taught about "intelligent design," a view of creation that challenges established scientific thinking and promotes the idea that an unseen force is behind the development of humanity. Although he said that curriculum decisions should be made by school districts rather than the federal government, Bush told Texas newspaper reporters in a group interview at the White House on Monday that he believes that intelligent design should be taught alongside evolution as competing theories.

Chapter 7: Betraying the Founders' Legacy: Democracy as a Weapon against Science

*A version of this chapter was published as "Rejecting the Founders' Legacy: Democracy as a Weapon against Science," in *Logos* 12, no. 2 (2013).

1. Chris Mooney, *The Republican War on Science* (New York: Basic Books, 2006), p. 35.
2. Ibid., p. 5.
3. Tanya Lewis, "A Year of Trump: Science Is a Major Casualty in the New Politics of Disruption," *Scientific American*, December 14, 2017, https://www.scientificamerican.com/article/a-year-of-trump-science-is-a-major-casualty-in-the-new-politics-of-disruption/. See also Jacob Carter, Gretchen Goldman, Genna Reed, et al., *Sidelining Science Since Day One: How the Trump Administration Has Harmed Public Health and Safety in Its First Six Months* (Cambridge, MA: Center for Science and Democracy, Union of Concerned Scientists, July 2017), https://www.ucsusa.org/center-science-and-democracy/promoting-scientific-integrity/sidelining-science-from-day-one#.WpMmmZPwZ24.
4. I. Bernard Cohen, *Science and the Founding Fathers: Science in the Political Thought of Thomas Jefferson, Benjamin Franklin, John Adams, and James Madison* (New York: W. W. Norton, 1995), p. 60.
5. Ibid., pp. 22–27.
6. Benjamin Franklin, *A Proposal for Promoting Useful Knowledge Among the British Plantations in America* (1743; Research Triangle Park, NC: National Humanities Center, 2009), p. 1, http://www.nationalhumanitiescenter.org/pds/becomingamer/ideas/text4/amerphilsociety.pdf.
7. Ibid.
8. Ibid., pp. 1–2.
9. I. Bernard Cohen, *Benjamin Franklin's Science* (Cambridge, MA: Harvard University Press, 1990), p. 7.
10. "George Washington Timeline," Washington Papers, University of Virginia, 2018, http://gwpapers.virginia.edu/history/biography-of-george-washington/.
11. George Washington, *Circular Letter of Farewell to the Army*, June 8, 1783, George Washington Collection, Library of Congress, http://www.loc.gov/teachers/classroommaterials/presentationsandactivities/presentations/timeline/amrev/peace/circular.html.
12. Richard Hofstadter, *Anti-Intellectualism in American Life* (New York: Alfred A. Knopf, 1963), p. 274.
13. George Washington, *Farewell Address*, 1796, Avalon Project, Yale Law School, http://avalon.law.yale.edu/18th_century/washing.asp.
14. Cohen, *Science and the Founding Fathers*, p. 196.
15. Ibid., p. 197.
16. John Adams, letter to Abigail Adams, May 12, 1780, Founders Online, National Archives, https://founders.archives.gov/documents/Adams/04-03-02-0258.
17. "History," American Academy of Arts and Sciences, 2018, https://www.amacad.org/content.aspx?i=7.
18. "Charter of Incorporation of the American Academy of Arts and Sciences," American Academy of Arts and Sciences, May 4, 1780, https://www.amacad.org/content/about/about.aspx?d=23.
19. Cohen, *Science and the Founding Fathers*, p. 63.

20. Ibid., p. 97.

21. Ibid., p. 67.

22. Thomas Jefferson, *Notes on the State of Virginia* (London: John Stockdale, 1787); available online on Google Books, https://books.google.com/books?id=UO0OAAAAQAAJ& pg=PP5#v=onepage&q&f=false.

23. Cohen, *Science and the Founding Fathers*, p. 121.

24. Ibid., p. 132.

25. Ibid., pp. 262–70.

26. Ibid., p. 60.

27. Barry Goldwater, "Quote for the Day," *Atlantic*, November 24, 2006, http://www .theatlantic.com/daily-dish/archive/2006/11/quote-for-the-day/232168/.

28. Mooney, *Republican War on Science*, p. 36.

29. "Blurred Lines," *Nature* 545 (May 11, 2017): 134, https://www.nature.com/polopoly _fs/1.21956!/menu/main/topColumns/topLeftColumn/pdf/545133b.pdf.

30. Michael D. Shear, "Huntsman on Evolution? 'Call Me Crazy,'" *Caucus* (blog), *New York Times*, August 18, 2011, https://thecaucus.blogs.nytimes.com/2011/08/18/huntsman -on-evolution-call-me-crazy/. See also Karoun Demirjian, "Senate Confirms Jon Huntsman as Russia Ambassador," *Washington Post*, September 28, 2017, https://www.washingtonpost.com/ powerpost/senate-confirms-jon-huntsman-as-russia-ambassador/2017/09/28/5bc2a6a4-a495 -11e7-ade1-76d061d56efa_story.html.

31. Matt Williams, "Republican Congressman Paul Broun Dismisses Evolution and Other Theories," *Guardian*, October 6, 2012, https://www.theguardian.com/world/2012/oct/06/ republican-congressman-paul-broun-evolution-video.

32. Jeffrey Mervis, "At House Science Panel Hearing, Sarcasm Rules," *Science*, March 28, 2014, http://www.sciencemag.org/news/2014/03/house-science-panel-hearing-sarcasm-rules.

33. "Early Life," Governor Bobby Jindal, September 4, 2012, https://web.archive.org/ web/20120904133312/http://www.bobbyjindal.com:80/news/about-bobby/88-early-life.

34. "RGA Announces New Leadership," Republican Governors Association, November 15, 2012, http://www.rga.org/homepage/rga-announces-new-leadership-2/.

35. Ray Nothstine, "6 Interesting Facts about Bobby Jindal's Christian Faith," *Christian Post*, July 7, 2015, https://www.christianpost.com/news/6-interesting-facts-about-bobby -jindals-christian-faith-141270/; Bill Barrow, "Science Law Could Set Tone for Jindal," *Times-Picayune*, June 26, 2008, http://www.nola.com/news/index.ssf/2008/06/science_law _could_set_tone_for.html; Xerxes A. Wilson, "Louisiana Outlaws Creation of Animal-Human Hybrids," LSU Now, July 16, 2009, http://www.lsunow.com/news/louisiana-outlaws-creation -of-animal-human-hybrids/article_93057c89-0a94-5aa8-8cb2-f3bb389c06ee.html.

36. Kenneth R. Miller, "Bobby Jindal's Science Problem," *Slate*, July 30, 2012, http://www .slate.com/articles/health_and_science/science/2012/07/bobby_jindal_possible_vice _presidential_pick_but_has_a_creationism_problem_.html.

37. *Encyclopedia of Religion*, s.v. "Theocracy," (New York: Macmillan Reference USA, 2005), p. 9110.

38. John Timmer, "Louisiana Passes First Antievolution 'Academic Freedom' Law," *Ars Technica*, June 27, 2008, https://arstechnica.com/tech-policy/2008/06/louisiana-passes -first-antievolution-academic-freedom-law/.

39. Edwards v. Aguillard, 482 U.S. 578 (1987), http://www.law.cornell.edu/supct/html/ historics/USSC_CR_0482_0578_ZO.html.

40. Louisiana Science Education Act (Act No. 473), S.B. 733, La. Rev. Stat. 17:285.1 (signed June 25, 2008), http://legis.la.gov/Legis/ViewDocument.aspx?d=503483.

41. "One on One with Governor Bobby Jindal," interview by Hoda Kotb, NBC News, April 12, 2013, video, 12:08, http://www.nbcnews.com/video/one-on-one-with-governor -bobby-jindal-26006595578, at 10:00–10:43.

42. Zack Kopplin, "Stop Governor Jindal's Creationist Voucher Program," OpEd News, July 16, 2012, http://www.opednews.com/Diary/Stop-Governor-Jindal-s-Cre-by -Zack-Kopplin-120716-764.html. See also Stephanie Simon, "Taxpayers Fund Teaching Creationism," *Politico*, March 24, 2014, https://www.politico.com/story/2014/03/education -creationism-104934; Stephanie Mencimer, "Mike Pence's Voucher Program in Indiana Was a Windfall for Religious Schools," *Mother Jones*, December 2, 2016, https://www.motherjones .com/politics/2016/12/mike-pence-voucher-program-religious-schools/.

43. Joshua Youngkin, "Dear Bill Moyers: An Open Letter," *Evolution News & Science Today*, March 8, 2013, http://www.evolutionnews.org/2013/03/dear_bill_moyer069921.html.

44. Ed Anderson, "Human-Animal Hybrid Ban Sought at Louisiana Session," *Times-Picayune*, April 17, 2009, http://www.nola.com/news/index.ssf/2009/04/humananimal _hybrid_ban_sought.html.

45. S.B. 115 (Act 108), La. Rev. Stat. 14:89.6 (signed June 19, 2009), http://legis.la.gov/ Legis/ViewDocument.aspx?d=664992.

46. "Green Light for Hybrid Research," BBC News, January 17, 2008, http://news.bbc .co.uk/2/hi/health/7193820.stm; Fergus Walsh, "UK's First Hybrid Embryos Created," BBC News, April 1, 2008, http://news.bbc.co.uk/2/hi/health/7323298.stm.

47. Daniel J. Loar and Robert M. Tasman, *Louisiana Conference of Catholic Bishops Legislative Update* 2, June 26, 2009, p. 2, http://www.laccb.org/files/2009_legislative-update_web.pdf.

48. Gene Mills, "2009 Session Finished!" *End of Week*, June 26, 2009 (on file with Barbara Forrest).

49. Rob Boston, "Perry Prayer-A-Palooza Panned," *Church & State*, September 2011, https://www.au.org/church-state/september-2011-church-state/featured/ perry-prayer-a-palooza-panned.

50. Margaret Johnson, Louisiana Right to Life, "State Rankings Place Louisiana as #1 Pro-Life State," news release, 2012, http://archive.constantcontact.com/fs076/1101796400807/ archive/1109106289423.html.

51. Charles Colson et al., "Evangelicals and Catholics Together: The Christian Mission in the Third Millennium," *First Things*, May 1994, https://www.firstthings.com/article/1994/05/ evangelicals-catholics-together-the-christian-mission-in-the-third-millennium; Robert George, Timothy George, and Chuck Colson, *Manhattan Declaration: A Call of Christian Conscience*, November 20, 2009, http://www.manhattandeclaration.org/. See also Frederick Clarkson, "Christian Right Seeks Renewal in Deepening Catholic-Protestant Alliance," Political Research Associates, July 23, 2013, https://www.politicalresearch.org/2013/07/23/ christian-right-seeks-renewal-in-deepening-catholic-protestant-alliance/.

52. Colson et al., "Evangelicals and Catholics Together."

53. Ibid.

54. George et al., *Manhattan Declaration*, p. 3.

55. Ibid., pp. 8–9.

56. Colson et al., "Evangelicals and Catholics Together."

57. Ibid.

58. Ibid.

59. See Barbara Forrest, "A Defense of Naturalism as a Defense of Secularism," in *Sidney Hook Reconsidered*, ed. Matthew J. Cotter (Amherst, NY: Prometheus Books, 2004); available at Internet Archive, https://web.archive.org/web/20110709203119/http://www .creationismstrojanhorse.com/Forrest_Defense_of_Naturalism.pdf.

60. Colson et al., "Evangelicals and Catholics Together."

61. George et al., *Manhattan Declaration*, p. 1.

62. Ibid., p. 7.

63. Ibid., p. 8.

64. Clarkson, "Christian Right Seeks Renewal."

65. Frederick Clarkson, "Dominionism Rising: A Theocratic Movement Hiding in Plain Sight," *Public Eye*, Summer 2016, p. 12, https://www.politicalresearch.org/wp-content/uploads/2016/10/PE_Summer2016_Clarkson.pdf.

66. Frederick Clarkson, "Remaking America as a Christian Nation," Political Research Associates, December 5, 2005, http://www.politicalresearch.org/2005/12/05/the-rise-of-dominionismremaking-america-as-a-christian-nation/#sthash.pPJrIqqH.dpbs.

67. Ibid.

68. Adam Nossiter, "In Louisiana, Inklings of a New (True) Champion of the Right," *New York Times*, June 2, 2008, http://www.nytimes.com/2008/06/02/us/02jindal.html.

69. "Mr. Theogene Anthony Mills," Lobbyist Registration Form, Louisiana Board of Ethics, 2017, http://ethics.la.gov/Lobbyist/upload/367/20161213_143409_367Year2017.pdf.

70. Articles of Incorporation of Louisiana Family Forum, September 15, 1997, Louisiana Secretary of State, on file with Barbara Forrest; "Tony Perkins, President," FRC Staff, Family Research Council, http://www.frc.org/tony-perkins. See also Jeremy Alford, "Holy Warriors," *Independent*, May 26, 2010, http://theind.com/article-6004-holy-warriors.html.

71. Kyle Mantyla, "Bobby Jindal's Prayer Rally Advocates Putting Christians in Control of Government and All Aspects of Society," Right Wing Watch, January 26, 2015, http://www.rightwingwatch.org/post/bobby-jindals-prayer-rally-advocates-putting-christians-in-control-of-government-and-all-aspects-of-society/; "RWW News: Gene Mills Preaches Seven Mountains Dominionism at Gov. Jindal's Prayer Rally," Right Wing Watch Blog, January 26, 2015, YouTube video, 2:47, https://www.youtube.com/watch?v=FbofEu2lFrg.

72. Clarkson, "Dominionism Rising," p. 13.

73. "RWW News: Gene Mills Preaches."

74. Kyle Mantyla, "Bobby Jindal Gets a Jump-Start on His Right-Wing Prayer Rally," Right Wing Watch, January 7, 2015, http://www.rightwingwatch.org/post/bobby-jindal-gets-a-jump-start-on-his-right-wing-prayer-rally/.

75. "US Public Becoming Less Religious," Pew Research Center, Washington, DC, November 3, 2015, p. 97, http://assets.pewresearch.org/wp-content/uploads/sites/11/2015/11/201.11.03_RLS_II_full_report.pdf; Gregory A. Smith and Jessica Martinez, "How the Faithful Voted: A Preliminary 2016 Analysis," Pew Research Center, Washington, DC, November 9, 2016, http://www.pewresearch.org/fact-tank/2016/11/09/how-the-faithful-voted-a-preliminary-2016-analysis/; "Religious Landscape Study: Party Affiliation," Pew Research Center, Washington, DC, 2014, http://www.pewforum.org/religious-landscape-study/party-affiliation/.

76. Alford, "Holy Warriors."

77. "Louisiana Governor Signs Creationist Bill," National Center for Science Education, June 27, 2008, https://ncse.com/news/2008/06/louisiana-governor-signs-creationist-bill-001437; Wallis Watkins, "Senator Ben Nevers Will Bring Legislative Experience to Chief of Staff," WRKF, November 25, 2015, http://wrkf.org/post/senator-ben-nevers-will-bring-legislative-experience-chief-staff; Melinda Deslatte, "Conservative Group Seeks to Sway La. Lawmakers," *Real Clear Politics*, July 30, 2011, http://www.realclearpolitics.com/news/ap/politics/2011/Jul/30/conservative_group_seeks_to_sway_la__lawmakers.html.

78. "Partisan Composition of State Legislatures 2002–2014," National Conference of State Legislatures, http://www.ncsl.org/documents/statevote/legiscontrol_2002_2014.pdf.

79. "2007 Governors' Christmas Gala," Louisiana Family Forum, video, December 2007, on file with Barbara Forrest.

80. "2017 Legislative Pastors' Briefing," Louisiana Family Forum, April 11, 2017, http://www.lafamilyforum.org/2017pastorsbriefing/.

81. "18th Annual Gala, 2017 LFF Legislative Awards," Louisiana Family Forum, 2017, http://www.lafamilyforum.org/2017gala/.

82. *2016 Louisiana Runoff Voter Guide*, Louisiana Family Forum, December 10, 2016, http://www.lafamilyforum.org/wp-content/uploads/2016/11/2016runoffguide-vFINAL.pdf; *2017 Legislative Scorecard*, Louisiana Family Forum Action Center, http://www.lafamilyforum.org/wp-content/uploads/2017Scorecard-1pgr-14in-corrected1.pdf.

83. Gene Mills, "Louisiana Family Forum 17th Annual Legislative Awards Gala—a Huge Success!" *End of Week*, Louisiana Family Forum, September 16, 2016, on file with Barbara Forrest. See also Gene Mills, "ULL LGBT . . . To Be Continued?" *End of Week*, Louisiana Family Forum, August 10, 2012, http://www.lafamilyforum.org/ull-lgbt-to-be-continued/.

84. Pearson Cross, "Cross Wise: Boxed In," *Independent*, September 2, 2016, http://theind.com/article-23851-cross-wise-boxed-in.html.

85. Gene Mills, "A Night of Honor," *End of Week*, Louisiana Family Forum, September 22, 2017, on file with Barbara Forrest; Congressman Steve Scalise, "Family Values," United States House of Representatives, https://scalise.house.gov/issues/family-values.

86. "Confirm Kyle Duncan to 5th Circuit!" *Family Facts*, Louisiana Family Forum, October 2017, on file with Barbara Forrest; Jeff Landry, "Trump Court Pick Kyle Duncan Is the Neil Gorsuch of Louisiana," *The Hill*, November 29, 2017, http://thehill.com/opinion/judiciary/362276-trump-circuit-court-pick-kyle-duncan-is-the-neil-gorsuch-of-louisiana.

87. Drew Broach, "Kyle Duncan Confirmed in Tight Senate Vote for 5th Circuit Judgeship," *Times-Picayune*, April 25, 2018, https://www.nola.com/national_politics/2018/04/kyle_duncan_judge_confirmed_senat.html.

88. Gene Mills, "National Prayer Breakfast—Washington, DC," *End of Week*, Louisiana Family Forum, February 9, 2018, on file with Barbara Forrest.

89. Gene Mills, "Kingdom Implications," *Solutions*, October 22, 2012, http://mysolutionsmagazine.com/kingdom-implications/.

90. "2007 Governors' Christmas Gala."

91. I am indebted to Louisiana resident and former Pentecostal pastor Jerry DeWitt, who explained the significance of this procedure. See Robert F. Worth, "From Bible Belt Pastor to Atheist Leader," *New York Times*, August 22, 2012, http://www.nytimes.com/2012/08/26/magazine/from-bible-belt-pastor-to-atheist-leader.html.

92. "About: Our Mission," Louisiana Family Forum, http://www.lafamilyforum.org/about/.

93. Gene Mills, "The New Louisiana," *End of Week*, newsletter, January 18, 2013, http://us2.campaign-archive1.com/?u=20dc9be01946aff7364f31092&id=2d3bb540f6&e=92108e1465; Governor Bobby Jindal, "Governor Jindal Orders June 27th as Statewide Day of Prayer for Perseverance Through Oil Spill Crisis," press release, June 24, 2010, http://emergency.louisiana.gov/Releases/06242010-prayer.html; Governor Bobby Jindal, *State of Louisiana Proclamation*, December 18, 2012, http://gallery.mailchimp.com/20dc9be01946aff7364f31092/files/JindalProclamation.pdf.

94. Clancy DuBos, "Da Winnas and da Loozas," *Gambit*, June 26, 2009, http://www.bestofneworleans.com/blogofneworleans/archives/2012/06/08/da-winnas-and-da-loozas.

95. Mills, "Kingdom Implications."

96. Ibid.

97. Ibid.

98. Sue Lincoln, "Take Me to Church: Gene Mills' Legislative Influence," WRKF, June 30, 2015, http://wrkf.org/post/take-me-church-gene-mills-legislative-influence.

99. Will Sentell, "New Science Standards Win Tentative State Approval, after Arguments over Evolution," *Advocate*, March 7, 2017, http://www.theadvocate.com/baton_rouge/news/education/article_1ff63ff0-02ab-11e7-9cdc-8b666e0646c4.html.

100. James Madison, "Detached Memoranda," *William and Mary Quarterly* 3, no. 4 (October 1946): 534, 554, 556, http://www.jstor.org/stable/1921903.

101. Thomas E. Mann and Norman J. Ornstein, "Let's Just Say It: The Republicans Are the Problem," *Washington Post*, April 27, 2012, http://articles.washingtonpost.com/2012-04-27/opinions/35453898_1_republican-party-party-moves-democratic-party.

Chapter 8: The Return of Determinism: Science, Power, and Sirens in Distress

1. The literal-minded miss Bierce's irony. For a good start on philosophy of science debates, see Imre Lakatos and Alan Musgrave, *Criticism and the Growth of Knowledge* (New York: Cambridge University Press, 1973).

2. Paul Feyerabend, *Against Method* (London: Verso, 1975). The book title should have been, "Against the Rule of Dogmatic Method," for he had nothing against methods—the more the merrier—just against the use of a single method or of methods unaccompanied by realistic humility. See Kurt Jacobsen, *Dead Reckonings: Ideas, Interests, and Politics in the "Information Age"* (Atlantic Highlands, NJ: Humanities Press, 1997), pp. 3–20.

3. Slavoj Žižek, *Violence* (London: Profile Books, 2009), pp. 69–70.

4. Quoted in ibid., p. 69. Feyerabend (*Against Method*) made a similar point.

5. As exemplified in the realm of public policy analysis, see John Kingdon's *Agendas, Alternatives, and Public Policies* (Boston: Little, Brown, 1984) and Kurt Jacobsen, *Chasing Progress in the Irish Republic* (Cambridge: Cambridge University Press, 1994).

6. The examples are legion. One is Robert Jervis's book on alleged Intelligence failure leading to the Iraq invasion: *Why Intelligence Fails: Lessons from the Iranian Revolution and the Iraq War* (Ithaca, NY: Cornell University Press, 2011). See also the retort by Fulton Armstrong, "The Damning Evidence: The CIA and WMDs," *New York Review of Books*, August 19, 2010.

7. Jane Mayer, *Dark Money* (London: Scribe, 2016), p. 6.

8. Chris Mooney, *The Republican War on Science* (New York: Basic Books, 2006).

9. See Theda Skocpol and Vanessa Williamson, *The Tea Party and the Remaking of Republican Conservatism* (New York: Oxford University Press, 2016); Jill Lepore, *The Whites of Their Eyes: The Tea Party's Revolution and the Battle over American History* (Princeton, NJ: Princeton University Press, 2010); Lauren Langman and George Lundskow, *God, Guns, Gold, and Glory: American Character and Its Discontents* (Leiden: Brill, 2016); Nancy MacLean, *Democracy in Chains* (New York: Viking, 2017); and Mayer, *Dark Money*.

10. The logic of business is coercion, monopoly, and destruction of the weak, not "choice" or "service" or "universal affluence." Thomas Franks, *One Market Under God* (London: Secker & Warburg, 2001), p. 87.

11. A nod here to Neil Sheehan's splendid *A Bright Shining Lie: John Paul Van and America in Vietnam* (New York: Random House, 1988).

12. Philip Mirowski, *Machine Dreams: Economics Becomes a Cyborg Science* (New York: Cambridge University Press, 2002).

13. James C. Scott, *Seeing Like a State: How Certain Schemes to Improve the Human Condition Have Failed* (New Haven, CT: Yale University Press, 1998).

14. Robert M. Young, "Science, Ideology, and Donna Haraway," *Science as Culture* 3, no. 2 (1992): 168.

15. Ibid.

16. "Their vision was of a society wholly made over in the image of the new mechanics—technically rationalized in every detail, predictable in every activity, and hence brought under total scientific management." Floyd Matson, *The Broken Image: Man, Science, and Society* (New York: Doubleday, 1966), p. 16.

17. Charles Sanders Peirce, "How to Make Our Ideas Clear," *Popular Science Monthly* January 1878, p. 237. See Jurgen Habermas's critique in his *Knowledge and Human Interests* (Boston: Beacon, 1972).

18. See David F. Noble, *The Religion of Technology: The Divinity of Man and the Spirit of Invention* (London: Penguin, 1999).

19. See Kurt Jacobsen, *International Politics and Inner Worlds: Masks of Reason under Scrutiny* (New York: Palgrave Macmillan, 2017), chap. 1.

20. Wendell Wallach, *A Dangerous Master* (New York: Basic Books, 2015), p. 28.

21. "Rational Knowledge does not controvert the tested findings of science; unlike empiricist philosophy however, it refuses to terminate with them." Max Horkheimer, *Critical Theory Selected Essays* (New York: Continuum, 1972), p. 164.

22. Jacques Ellul, *The Technological Society* (New York: Vintage, 1964).

23. Barry Richards, *Images of Freud: Cultural Responses to Psychoanalysis* (London: Dent, 1989), p. 3.

24. Lewis Mumford, *The Myth of the Machine*, 2 vols. (New York: Harcourt, Brace, Jovanovich, 1967–70); David F. Noble, *Forces of Production: A Social History of Industrial Automation* (New York: Knopf, 1990); Langdon Winner, *Autonomous Technology: Technics-Out-of-Control as a Theme in Political Thought* (Cambridge, MA: MIT Press, 1977); Reinhard Skinner "Technological Determinism: A Critique of Convergence Theory," *Comparative Studies in Society and History* 18, no. 1 (January 1976); Horkheimer, *Critical Theory*; Siegfried Giedeon, *Mechanization Takes Command: A Contribution to Anonymous History* (New York: Oxford University Press, 1948); Robert Heilbroner, "Do Machines Make History?" *Technology and Culture* 8, no. 3 (1967); W. E. Bijker, T. P. Hughes, and Trevor Pinch, *The Social Construction of Technological Systems: New Directions in the Sociology and History of Technology* (Cambridge, MA: MIT Press, 1987); Herbert Marcuse, *One Dimensional Man: Studies in the Ideology of Advanced Industrial Society* (Boston: Beacon Press, 2014); M. R. Smith and Leo Marx, eds., *Does Technology Drive History? Dilemmas of Technological Determinism* (Cambridge, MA: MIT Press, 1994); Kurt Jacobsen, *Technical Fouls: Democratic Dilemmas and Technological Change* (Boulder: Westview Press, 2000); David Edgerton, "From Innovation to Use: Ten Eclectic Theses on the Historiography of Technology," *History and Technology* 16 (1999); Theodore Roszak, *The Making of a Counter-Culture* (New York: Doubleday, 1969); and Jurgen Habermas, "Technology and Science as Ideology," in *Toward a Rational Society* (London: Heinemann, 1971).

25. See Langdon Winner, "Where Technological Determinism Went," in *Visions of STS*, ed. Stephen. H. Cutliffe and Carl Mitcham (Albany: SUNY Press, 2001), p. 14; Paul Ceruzzi, "Moore's Law and Technological Determinism: Reflections on the History of Technology," *Technology and Culture* 46, no. 3 (July 2005); Sally Wyatt, "Technological Determinism Is Dead, Long Live Technological Determinism," in *The Handbook of Science and Technology Studies*, ed. Edward J. Hackett, Olga Amsterdamska, Michael Lynch, and Judy Wacjman (Cambridge, MA: MIT Press, 2008); Cyrus C. M. Mody, "Small but Determined: Technological Determinism in Nanoscience," *International Journal for Philosophy of Chemistry* 10, no. 2 (2004); and Taylor Dotson, "Technological Determinism and Permissionless Innovation as Technocratic Governing Mentalities," *Engaging Science, Technology, and Society* 1 (2015).

26. R. D. Laing, *The Divided Self* (London: Penguin, 1965).

27. Hannah Arendt, *The Human Condition* (Chicago: University of Chicago Press, 1960), pp. 322–23.

28. Benjamin Page and Martin Gilens, *Democracy in America? What Has Gone Wrong and What We Can Do About It* (Chicago: University of Chicago Press, 2017).

29. Wallach, *Dangerous Master*, pp. 45, 50–51; and Eugene Schwartz, *Overskill: The Decline of Technology in Modern Civilization* (New York: Times Books, 1971).

30. George Monbiot, "The Corporate Stooges Who Nobble Serious Science," *Guardian*, February 24, 2004; Sheila Jasanoff, "Science, Politics, and the Renegotiation of Expertise at

EPA," *Osris* 7 (1992): 194–217, found experts recruited to back politician cases, leading to a decline in public belief in the scientist ability to "speak truth to power." Dorothy Nelkin, *Selling Science: How the Press Covers Science and Technology*, 2nd ed. (New York: W. H. Freeman, 1995); John P. Ioannidis, "Why Most Research Findings Are False," *PLoS Med* 2, no. 8 (2005).

31. Mooney, *Republican War on Science*, p. 8.

32. Ibid. p. 9.

33. David Smail, *Why Therapy Doesn't Work and What We Should Do About It* (New York: Robinson, 2001), p. 97.

34. British Medical Association, *Human Genetics: Choice and Responsibility* (Oxford: Oxford University Press, 1998), p. 3.

35. Daniel Gasman, *The Scientific Origins of National Socialism* (New Brunswick, NJ: Transaction Books, 1971), p. 91. Also see Steven Weiss, *The Nazi Symbiosis: Human Genetics and Politics in the Third Reich* (Chicago: University of Chicago Press, 2010); Benno Muller-Hill, *Murderous Science* (Plainview, NY: Cold Spring Harbor Laboratory Press, 1988).

36. Michael Sherry, *The Rise of American Air Power* (New Haven, CT: Yale University Press, 1987), p. 54.

37. Max Weber, *The Protestant Ethic and the Spirit of Capitalism* (New York: Scribner, 1930).

38. Mahmood Mamdani, *When Victims Become Killers: Colonialism, Nativism and Genocide in Rwanda* (Princeton, NJ: Princeton University Press, 2001), p. 77.

39. Evelyn Fox Keller, *The Century of the Gene* (Cambridge, MA: Harvard University Press 2000), p. 112. "Just how many other players—including regulatory sequences found elsewhere on the genome, the products of many other structural and regulatory genes, the complex signaling network of the living cell—are organized into a well-functioning and reliable whole is the question that dominates the attention of molecular biologists today," p. 72.

40. *Time* magazine in the 1990s noted that most people oppose human genetic engineering for any purpose except to cure diseases—which provides a huge exception. Philip Elmer-Dewitt, "The Genetic Revolution," *Time* January 17, 1994.

41. Jeremy Gruber, "The Unfulfilled Promise of Genomics," in *Genetic Explanations: Sense and Nonsense*, ed. Sheldon Krimsky, and Jeremy Gruber (Cambridge, MA: Harvard University Press, 2013).

42. Steve Jones, "Darwinism and Genes" in *What Scientists Think*, ed. Jeremy Stangroom (London: Routledge, 2005), p. 19.

43. Roar Fosse, Jay Joseph, and Mike Jones, "Schizophrenia: A Critical View on Genetic Effects," *Psychosis* 8, no. 1 (2016): 9.

44. David Plomin, "Child Development and Molecular Genetics," *Child Development* 84, no. 1 (January/February 2013): 104.

45. See Jonathan Leo, "The Search for Schizophrenia Genes," *Issues in Science & Technology* (Winter 2016): 88.

46. Jay Joseph and Claudia Chaufan, "Missing Heritability of Common Disorders: Should Health Researchers Care?" *International Journal of Health Services* 43, no. 2 (2013): 285n3.

47. Ibid., p. 289.

48. Jo C. Phelan, "Geneticization of Deviant Behavior and Consequences for Stigma: The Case of Mental Illness," *Journal of Health and Social Behavior* 46 (2005).

49. Alvin A Rosenfeld, "Child and Adolescent Mental Disorders Research: Current Directions, Future Needs," *Archives of General Psychiatry* 52, no. 9 (September 1995): 731.

50. David Bell, "The Power-Point Philosophe," *Nation*, March 7, 2018.

51. Ruth Hubbard, "The Mismeasure of the Gene," in *Genetic Explanations*, ed. Krimsky and Gruber, p. 19.

52. See Stuart Newman, "Evolution Is Not Mainly a Matter of Genes," in *Genetic Explanations*, ed. Krimsky and Gruber, pp. 26–33.

53. David S. Moore, "Big B Little b: Myth No. 1 Is That Mendelian Genes Actually Exist," in *Genetic Explanations*, ed. Krimsky and Gruber, pp. 43–50.

54. Katherine Hignett, "Scott Kelly: A NASA Twins Study Confirms Astronaut's DNA Actually Changed in Space," *Newsweek*, March 9, 2018; Erin Brodwin, "NASA Sent an Astronaut into Space for a Year—and It May Have Permanently Changed 7% of His DNA," *Business Insider*, March 8, 2018.

55. Evelyn Fox Keller, "Genes as Difference Makers," in *Genetic Explanations*, ed. Krimsky and Gruber, p. 41.

56. Evan Charney, "Review Essay on Gruber and Krimsky," *Logos* 12, no. 3 (Spring 2013). Also see Charney, "Politics, Genetics, and 'Greedy Reductionism,'" *Perspectives on Politics* 6, no. 2 (June 2008): 337–43; and Charney, "Behavior Genetics and Postgenomics," *Behavioral and Brain Sciences* (2012): 35, 6.

57. Richard Lewontin, Steven Rose, and Leon Kamin, *Not in Our Genes: Biology, Ideology and Human Nature* (New York: Pantheon Books, 1984); D. D. Jackson, "A Critique of the Literature on the Genetics of Schizophrenia," in *The Etiology of Schizophrenia*, ed. D. D. Jackson (New York: Basic Books, 1960); and Jay Joseph, *The Missing Gene: Psychiatry, Heredity, and the Fruitless Search for Genes* (New York: Algora, 2006).

58. Hervey Cleckley, *The Mask of Sanity: An Attempt to Reinterpret the So-Called Psychopathic Personality* (St. Louis: Mosby, 1941), p. 467.

59. Thomas Kuhn, *The Structure of Scientific Revolutions* (Chicago: University of Chicago Press, 1962).

60. Jonathan Leo, "Memo to the Newest Generation of Gene Hunters: Read Jay Joseph," in *Psychosis* 10, no. 1 (January 2018): 2.

61. See the critique of these "echo chambers" in Kurt Jacobsen, "Much Ado about Ideas: The Cognitive Factor in Economic Policy," *World Politics* 47, no. 2 (January 1995): 283–310.

62. Michael J. Joyner, Nigel Paneth, and John P. A. Ioannidis, "What Happens When Underperforming Big Ideas in Research Become Entrenched?" *Journal of the American Medical Association* 316, no. 13 (October 2016): 1355.

63. On Lysenkoism see David Joravsky, *The Lysenko Affair* (Chicago: University of Chicago Press, 1970).

64. Sarah Boseley, "Pharmaceutical Adverts 'Inadequately Policed,'" *Guardian*, September 23, 2003, p. 10.

65. See David I. Harvie, *Limeys: The Conquest of Scurvy* (London: Sutton Books, 2005).

66. Maya Salam, "The Opioid Epidemic: A Crisis in the Making," *New York Times*, October 26, 2017.

67. Steve Connor, "Glaxo Chief: Our Drugs Do Not Work on Most Patients," *Independent*, December 8, 2003.

68. Ibid.

69. Sidney Wolfe, "Worst Pills, Best Pills," *EXTRA* 14, no. 2 (March/April 2001): 11.

70. John Read, Olga Runciman, and Jacqui Dillon, "In Search of an Evidence Based Role for Psychiatry," *Future Science OA* 2, no. 1 (March 2016):. 1.

71. Ibid. See also John Read, Lorenza Magliano, and Vanessa Beavan, "Public Beliefs about the Causes of 'Schizophrenia': Bad Things Happen and Can Drive You Crazy," in *Models of Madness: Psychological, Social, and Biological Approaches to Psychosis*, ed. John Read and Jacqui Dillon (London: Routledge, 2014).

72. Connor, "Glaxo Chief."

73. "Prescription Drugs: 7 out of 10 Americans Take at Least One, Study Finds," *HuffPost*, June 20, 2013.

74. Marcia Angell, *The Truth about Drug Companies: How They Deceive Us and What to Do About It* (New York: Random House, 2005).

75. Sara G. Miller, "Drug Use in America: What the Numbers Say," *Live Science* September 8, 2016.

76. Peter Gay, *Reading Freud: Explorations and Entertainments* (New Haven, CT: Yale University Press, 1990), p. 83.

77. Lewis Lapham, "Bomb-o-Gram," in *Waiting for the Barbarians* (London: Verso, 1997), p. 30.

78. Kurt Jacobsen, *Pacification and Its Discontents* (Chicago: Prickly Paradigm Press, 2009).

79. Zalin Grant, *Facing the Phoenix: The CIA and the Political Defeat of the United States in Vietnam* (New York: Norton, 1991); Mark Moyar, *Phoenix and the Birds of Prey: The CIA's Secret Campaign to Destroy the Viet Cong* (Annapolis: Naval Institute Press, 1997); Lewis Sorley, *A Better War: The Unexamined Victories and Final Tragedy of America's Last Years in Vietnam* (New York: Harcourt Brace, 1999); and James S. Robbins, *This Time We Win: Revisiting the Tet Offensive* (New York: Encounter Books, 2010).

80. General William DePuy, "Vietnam: What We Might Have Done and Why We Didn't Do It," *Army* 36, no. 2 (February 1986): 81.

81. Vietnam Veterans Against the War, *The Winter Soldier Investigations: An Inquiry into American War Crimes* (Boston: Beacon Press, 1972; transcript of event held in Detroit, January 31–February 2 1971), pp. 3, 39.

82. William Colby, *Lost Victory: A Firsthand Account of America's Sixteen-Year Involvement in Vietnam* (New York: McGraw-Hill, 1989), p. 25.

83. David Elliott, "Parallel Wars? Can 'Lessons of Vietnam' Be Applied to Iraq?" in *Iraq and The Lessons of Vietnam: Or, How Not to Learn from the Past*, ed. Lloyd Gardner and Marilyn Young (New York: New Press, 2008).

84. George W. Allen, *None So Blind: A Personal Account of the Intelligence Failure in Vietnam* (Chicago: Ivan R Dee, 2001), pp. 260, 265.

85. Gabriel Kolko, "The Political Significance of the Center for Vietnamese Studies and Programs," *Bulletin of Concerned Asian Scholars* 3, no. 2 (February 1971): 42.

86. Stathis Kalyvas and Matthew Kocher, "Dynamics of Violence in Civil War: Evidence from Vietnam," (working paper, Yale University, New Haven, CT, 2006), p. 9.

87. General Cao Van Vien and Lt. General Dong Van Khuyen, "*Reflections on the Vietnam War*" (Washington, DC: US Army Center of Military History, 1980), p. 68.

88. James C. Scott and Matthew Light, "The Misuse of Numbers: Audits, Quantification, and the Obfuscation of Politics," in *Reconsidering American Power*, ed. John D. Kelly, Kurt Jacobsen and Marston Morgan (Oxford: Oxford University Press, 2019).

89. Ibid.

90. Anders Sweetland, *Item Analysis of the HES (Hamlet Evaluation System)* (Santa Monica: RAND Corporation, 1968), p. 1. He cites one aggrieved critic saying, "The HES is no damn good, it didn't predict Tet."

91. Austin Long, *On "Other War": Lessons from Five Decades of RAND Counterinsurgency Research* (Santa Monica: RAND, 2006), p. 40.

92. David W. P. Elliot, *The Vietnamese War: Revolution and Social Change in the Mekong Delta, 1930–1975*, vol. 2 (Armonk, NY: M. E. Sharpe, 2003).

93. Ibid., p. 1211.

94. Ronald Spector, *After Tet: The Bloodiest Year in Vietnam* (New York: Free Press, 1993), p. 293.

95. David F. Schmitz, *The Tet Offensive: Politics, War, and Public Opinion* (Lanham: Rowman & Littlefield, 2005), p. 37.

96. David Hunt, *Vietnam's Southern Revolution: From Peasant Insurrection to Total War, 1959–1968* (Amherst: University of Massachusetts Press, 2008), p. 355.

97. Dave Young, "Computing War Narratives: The Hamlet Evaluation System in

Vietnam," *APRJA (A Peer-Reviewed Journal About)* 6, no. 1 (2017): 14, http://www.aprja.net/computing-war-narratives-the-hamlet-evaluation-system-in-vietnam/?pdf=3204.

98. In this vein behold Stathis N. Kalyvas and Matthew Adam Kocher, "The Dynamics of Violence in Vietnam: An Analysis of the Hamlet Evaluation System," *Journal of Peace Research* 46, no. 3 (May 2009) and Matthew Kocher, Thomas B. Pepinsky, and Stathis Kalyvas, "Aerial Bombardment, Indiscriminate Violence, and Territorial Control in Unconventional Wars: Evidence from Vietnam," *American Journal of Political Science* 55, no. 2 (April 2011).

99. See Hugh Gusterson, *Drone: Remote Control Warfare* (Cambridge: MIT Press 2016); Grigoire Chamayou, *A Theory of the Drone* (New York: New Press, 2013); Medea Benjamin, *Drone Warfare: Killing by Remote Control* (New York: Verso, 2013); Andrew Cockburn, *Kill Chain: The Rise of the High-Tech Assassins* (New York: Verso, 2016); P. W. Singer, *Wired for War: The Robotics Revolution and Conflict in the 21st Century* (New York: Penguin, 2011); and Jeremy Scahill, *The Assassination Complex: Inside the Government's Secret Drone Warfare Program* (New York: Simon & Schuster, 2017).

100. Gusterson, *Drone*, pp. 23–24.

101. Jon Boone, "US Drone Strikes Could Be Classified as War Crimes, Says Amnesty International," *Guardian*, October 22, 2013.

102. Robert M. Young, *Mind, Brain, and Adaptation in the Nineteenth Century: Cerebral Localization and Its Biological Context from Gall to Ferrier* (Oxford: Clarendon, 1970), p. 31.

103. Brian Charlesworth and Deborah Charlesworth, "Geneticists Know There's More to Life," *Guardian* March 11, 2018.

104. Nicholas Rescher, *The Limits of Science*, 2nd ed. (Pittsburgh: University of Pittsburgh Press, 1999), p. 35.

105. Noam Chomsky, *For Reasons of State* (New York: Pantheon, 1973), p. 42.

106. Curtis Bowman, "Odysseus and the Siren Call of Reason: The Frankfurt School Critique of Enlightenment," *Other Voices: The (e)journal of Cultural Criticism* 1, no. 1 (March 1997).

107. Nancy Love, "Why Do the Sirens Sing? Figuring the Feminine in Dialectic of Enlightenment," in *Rethinking the Frankfurt School: Alternative Legacies of Cultural Critique*, ed. Jeffrey T. Nealon and Carn Irr (Albany: State University of New York Press, 2002).

108. Objective reason, which rejects stunted formal theories, means criticism and by "criticism, we mean the intellectual, and eventually practical effort which is not satisfied to accept the prevailing ideas, actions, and social conditions unthinkingly and from mere habit; effort which aims to coordinate the individual sides of social life with each other and with the general ideas and aims of the epoch, to deduce them genetically, to distinguish the appearance from essence, to examine the foundation of things, in short, really to know them." Horkheimer, *Critical Theory: Selected Essays*, p. 270.

Chapter 9: Back to the Futurists: On Accelerationism Left and Right

1. Elizabeth Dias, "What You Missed While Not Watching the Bill Nye and Ken Ham Creation Debate," *Time*, February 5, 2014, http://time.com/4511/bill-nye-ken-ham-debate/.

2. Mark C. Biedebach, "Atheism Comes into the Classroom through the Back Door," chap. 3 in *Evolution vs. Creation . . . a New Approach to Teaching How Life Began* (manuscript in progress; California State University, Long Beach, CA), http://web.csulb.edu/~mbiedeba/ch3.html.

3. Nick Land, "The Dark Enlightenment," *Dark Enlightenment* (blog), December 25, 2012, http://www.thedarkenlightenment.com/the-dark-enlightenment-by-nick-land/; Nick Srnicek and Alex Williams, "#Accelerate: Manifesto for an Accelerationist Politics," in

#Accelerate: The Accelerationist Reader, ed. Robin Mackay and Armen Avanessian, 2nd ed. (Falmouth, UK: Urbanomic Media, 2017), pp. 347–62; Laboria Cuboniks, "Xenofeminism: A Politics for Alienation," *Xenofeminism*, June 11, 2015, http://www.laboriacuboniks.net/qx8bq .txt; "#AltWoke Manifesto," *&&& Journal*, February 5, 2017, http://tripleampersand.org/alt -woke-manifesto/; Nick Srnicek and Alex Williams, *Inventing the Future: Postcapitalism and a World without Work* (London: Verso, 2015).

4. Andy Beckett, "Accelerationism: How a Fringe Philosophy Predicted the Future We Live in," *Guardian*, May 11, 2017, http://www.theguardian.com/world/2017/may/11/ accelerationism-how-a-fringe-philosophy-predicted-the-future-we-live-in.

5. It is true that many accelerationist writers embrace the rhetoric of modernity, and even the Enlightenment. See, for example, Srnicek and Williams, *Inventing the Future*, p. 49. Nonetheless, these writers consistently eschew the major conceptual tenets of radical Enlightenment thought, namely, the existence of an intelligible, deterministic, universe governed by equally intelligible and stable natural laws.

6. Nick Land, *The Thirst for Annihilation: Georges Bataille and Virulent Nihilism* (London: Routledge, 2002), pp. 1–9.

7. Manjit Kumar, *Quantum: Einstein, Bohr, and the Great Debate about the Nature of Reality* (repr.; New York: W. W. Norton, 2011), p. 331.

8. Alan Ramon Clinton, *Mechanical Occult: Automatism, Modernism, and the Specter of Politics* (New York: Peter Lang, International Academic Publishers, 2004), p. 193.

9. Friedrich Nietzsche, *The Gay Science: With a Prelude in Rhymes and an Appendix of Songs*, trans. Walter Kaufmann (New York: Vintage Books, 1974), p. 181.

10. Robert C. Holub, *Nietzsche's Jewish Problem: Between Anti-Semitism and Anti-Judaism* (Princeton, NJ: Princeton University Press, 2016), pp. 166–203.

11. Don Dombowsky, *Nietzsche's Machiavellian Politics* (Basingstoke, UK: Palgrave Macmillan, 2004), pp. 9–66.

12. Frank Cameron and Don Dombowsky, eds., *Political Writings of Friedrich Nietzsche: An Edited Anthology* (London: Palgrave Macmillan, 2008), p. 236.

13. Friedrich Nietzsche, *Thus Spoke Zarathustra: A Book for All and None*, ed. Adrian Del Caro and Robert B. Pippin, trans. Adrian Del Caro (New York: Cambridge University Press, 2006), p. 10.

14. Certainly, there were other proponents of Italian futurism, some of whom even occupied the left wing of the political spectrum. However, at least in Italy, futurism did come to be dominated by a right-wing and nationalist ethos, and Marinetti himself was instrumental in this.

15. Filippo Tommaso Marinetti, "To My Pegasus," in *Selected Poems and Related Prose*, ed. Luce Marinetti, trans. Elizabeth R. Napier and Barbara R. Studholme (New Haven, CT: Yale University Press, 2002), p. 38.

16. Filippo Marinetti, quoted in, Walter Benjamin, "The Work of Art in the Age of Mechanical Reproduction," in *Illuminations*, ed. Hannah Arendt, trans. Harry Zohn (New York: Schocken Books, 1969), pp. 19–20.

17. Ernst Jünger, "Total Mobilization," in *The Heidegger Controversy: A Critical Reader*, ed. Richard Wolin (Cambridge, MA: MIT Press, 1993), pp. 126–27.

18. Ibid., p. 129.

19. Ibid., p. 128.

20. For the will is "world-forming," rather than formed by a common world. See Ernst Jünger, *The Worker: Dominion and Form*, ed. Laurence Paul Hemming, trans. Bogdan Costea and Laurence Paul Hemming (Evanston, IL: Northwestern University Press, 2017), pp. 189–90. Commentators on Jünger have likewise pointed out the extent of his valorization of the

particular will, and how this fundamental worldview could accommodate the anti-Semitism of his time. While opposing the alarmism of Nazi propaganda, Jünger did identify the liberal Jew as nonetheless "alien" to German culture, and so incapable of "playing a creative role" in German life." See Thomas R. Nevin, *Ernst Jünger and Germany: Into the Abyss, 1914–1945* (Durham, NC: Duke University Press, 1996), pp. 74, 109.

21. Jünger, *Worker*, p. xv.

22. Oswald Spengler, *Man and Technics: A Contribution to a Philosophy of Life*, trans. Charles Francis Atkinson (1932; New York: Routledge, 2017), p. 82.

23. Ibid.

24. Benjamin Noys, "Futures of Accelerationism" (seminar, Faster/Slower/Future, Towards Postcapitalism, Kaaitheater, Brussels, Belgium, October 22, 2016).

25. Spengler, *Man and Technics*, p. 87.

26. Ibid., p. 86.

27. Ibid.

28. C. L. R. James, *Every Cook Can Govern: A Study of Democracy in Ancient Greece It's Meaning for Today*, 2nd ed. (Detroit, MI: Bewick Editions, 1992).

29. Karl Marx and Friedrich Engels, "Manifesto of the Communist Party," in *Marx & Engels Collected Works*, vol. 6 (London: Lawrence and Wishart, 2010), p. 487.

30. Martin Heidegger, *Introduction to Metaphysics*, trans. Gregory Fried and Richard Polt (New Haven, CT: Yale University Press, 2000), p. 213.

31. Friedrich Hölderlin, as quoted in Martin Heidegger, "The Question Concerning Technology," in *The Question Concerning Technology and Other Essays*, trans. William Lovitt (New York: Garland, 1977), p. 28.

32. Richard Wolin, ed., "'Only a God Can Save Us': *Der Spiegel*'s Interview with Martin Heidegger," in *Heidegger Controversy*, p. 91.

33. This account of French accelerationism is heavily indebted to Benjamin Noys's two books on the subject, *The Persistence of the Negative: A Critique of Contemporary Continental Theory* (Edinburgh: Edinburgh University Press, 2010), p. 5, and *Malign Velocities: Accelerationism and Capitalism* (Winchester, UK: Zero Books, 2014), p. xi. It was Noys who reintroduced the term "accelerationism" into the modern lexicon, and, further, identified accelerationist thought as bound up with the conditions of late capitalism.

34. Karl Marx, *Capital: Volume III*, trans. David Fernbach (London: Penguin Classics, 1991), as quoted in, Noys, *Malign Velocities*, pp. 8–10.

35. Marx, *Capital*, p. 959.

36. Gilles Deleuze and Félix Guattari, *Anti-Oedipus: Capitalism and Schizophrenia*, trans. Robert Hurley, Mark Seem, and Helen R. Lane (Minneapolis: University of Minnesota Press, 1983), pp. 239–40; Noys, *Malign Velocities*, pp. 1–2.

37. Deleuze and Guattari, *Anti-Oedipus*, p. 240.

38. Noys, *Malign Velocities*, p. 4.

39. Jean-François Lyotard, *Libidinal Economy*, trans. Iain Hamilton Grant (Bloomington: Indiana University Press, 1993), p. 111; cf. Noys, *Malign Velocities*, p. 3.

40. Jean Baudrillard, *Symbolic Exchange and Death*, trans. Iain Hamilton Grant, rev. ed. (London: SAGE Publications, 2017), p. 58.

41. Jean Baudrillard, "When Bataille Attacked the Metaphysical Principle of Economy," trans. David James Miller, *Canadian Journal of Political and Social Theory* 11, no. 3 (1987): 60.

42. Nick Land, *Fanged Noumena: Collected Writings 1987–2007*, ed. Ray Brassier and Robin Mackay, 2nd ed. (Falmouth, UK: Urbanomic, 2012), p. 446.

43. Noys, *Malign Velocities*, p. 54.

44. Land, *Fanged Noumena*, p. 21.

45. Karl Popper, *Quantum Theory and the Schism in Physics: From the Postscript to the Logic of Scientific Discovery*, ed. W. W. Bartley III, 1st ed. (London: Routledge, 1992), p. 175.

46. Eric Oberheim and Paul Hoyningen-Huene, "The Incommensurability of Scientific Theories," *The Stanford Encyclopedia of Philosophy*, ed. Edward N. Zalta, last updated March 5, 2013, https://plato.stanford.edu/archives/win2016/entries/incommensurability/.

47. Land, *Fanged Noumena*, p. 591; cf. Paul Feyerabend, *Against Method*, new ed. (London: Verso, 2010), p. 14.

48. Land, *Fanged Noumena*, p. 592.

49. Mike Riddle, "Doesn't Carbon-14 Dating Disprove the Bible?" in *The New Answers Book: Over 25 Questions on Creation/Evolution and the Bible*, ed. Ken Ham (Green Forest, AR: Master Books, 2006), pp. 77–87.

50. Paul Feyerabend, *Knowledge, Science, and Relativism* (Cambridge, UK: Cambridge University Press, 1999), p. 183.

51. In this way, science collapses into technology. There can be no sense of knowledge acquisition for its own sake, i.e., a pure theoretical attitude. Rather, all scientific pursuits are instrumentalized for specific ends and involve invention. Hence the rise of what is now termed "technoscience."

52. Land, "Dark Enlightenment."

53. Nick Land, "IQ Shredders," *Outside In: Involvements with Reality* (blog), July 17, 2014, http://www.xenosystems.net/iq-shredders/.

54. See Land's critique of the "Labor Theory of Value," in *Fanged Noumena*, pp. 346–47.

55. On the Left-Nietzscheanism of Russian futurism, see Bernice Glatzer Rosenthal, *New Myth, New World: From Nietzsche to Stalinism*, 1st ed. (University Park, PA: Penn State University Press, 2002); on the contradictions of Russian futurism, see Leon Trotsky, *Literature and Revolution*, ed. William Keach (Chicago, IL: Haymarket Books, 2005).

56. Many of these are anthologized by Robin Mackay and Armen Avanessian, eds., *#Accelerate: The Accelerationist Reader*, 2nd ed. (Falmouth, UK: Urbanomic Media, 2017); As for the art scene, one example of this is the LD50 Gallery, which was heavily criticized for promoting Far-Right figures. See "Why Is Nick Land Still Embraced by Segments of the British Art and Theory Scenes?" *E-Flux Conversations* (blog), March 17, 2017, https://conversations.e-flux.com/t/why-is-nick-land-still-embraced-by-segments-of-the-british-art-and-theory-scenes/6329.

57. Srnicek and Williams, "#Accelerate: Manifesto," pp. 351–52.

58. Srnicek and Williams, *Inventing the Future*, p. 57.

59. Ibid., p. 82.

60. Ibid., p. 83.

61. Ibid., p. 82.

62. Sadie Plant, "Binary Sexes, Binary Codes," as quoted in, Srnicek and Williams, *Inventing the Future*, p. 82.

63. Srnicek and Williams, *Inventing the Future*, pp. 76, 83n74.

64. Ibid., p. 78.

65. Ibid., p. 82.

66. Jon Swaine, "Donald Trump's Team Defends 'Alternative Facts' after Widespread Protests," *Guardian*, January 23, 2017, http://www.theguardian.com/us-news/2017/jan/22/donald-trump-kellyanne-conway-inauguration-alternative-facts.

67. Donald J. Trump and Tony Schwartz, *Trump: The Art of the Deal*, 1st ed. (New York: Random House, 1987), p. 58.

68. Norman Vincent Peale gave sermons at Marble Collegiate Church in Manhattan. Fred Trump would take his family on Sundays to listen to Peale's sermons on the "power of positive thinking." Peale also officiated at Donald Trump's wedding to Ivana Zelníčková. See David Brody and Scott Lamb, *The Faith of Donald J. Trump: A Spiritual Biography* (New York: Broadside Books, 2018).

69. Michael D'Antonio, *The Truth about Trump* (New York: Thomas Dunne Books, 2015), p. 39; Norman Vincent Peale, *The Power of Positive Thinking* (New York: Fireside,

2003), p. 2. Of course, Peale's phrase is taken from Philippians 4:13. However, in Peale's hands, these words are turned into a practical "formula" or means of "self-hypnosis" for overcoming all worldly obstacles, if only they are said out loud.

Chapter 10: The Myth of the Expert as Elite: Postmodern Theory, Right-Wing Populism, and the Assault on Truth

1. Donald Trump, quoted in Nick Gass, "Trump: 'The Experts Are Terrible,'" *Politico*, April 4, 2016, https://www.politico.com/blogs/2016-gop-primary-live-updates-and-results/2016/04/donald-trump-foreign-policy-experts-221528.

2. Michael Gove quoted in Henry Mance, "Britain Has Had Enough of Experts, Says Gove," *Financial Times*, June 3, 2016.

3. Wolfgang Streeck, "The Return of the Repressed as the Beginning of the End of Neoliberal Capitalism" in *The Great Regression*, ed. Heinrich Geiselberger (Cambridge: Polity Press, 2017), pp. 159–60.

4. Daniel C. Dennett, "The Hoax of Intelligent Design and How It Was Perpetrated," in *Intelligent Thought: Science versus the Intelligent Design Movement*, ed. John Brockman (New York: Vintage Books, 2006), p. 34.

5. There is now a rich literature on the global rise of populism. Some of the best works include Jan-Werner Müller, *What Is Populism?* (Philadelphia: University of Pennsylvania Press, 2016); John B. Judis, *The Populist Explosion: How the Great Recession Transformed American and European Politics* (New York: Columbia Global Reports, 2016); Cas Mudde and Cristóbal Rovira Kaltwasser, *Populism: A Very Short Introduction* (Oxford: Oxford University Press, 2017); Benjamin Moffitt, *The Global Rise of Populism: Performance, Political Style, and Representation* (Stanford: Stanford University Press, 2016); and Federico Finchelstein, *From Fascism to Populism in History* (Berkeley: University of California Press, 2017).

6. The anthropologist Angela Nagle has made a similar point in her *Kill All Normies: Online Culture Wars from 4chan and Tumblr to Trump and the Alt-Right* (Winchester, UK: Zero Books, 2017), p. 62. Shawn Otto has made a somewhat similar claim about a connection between postmodern thought and neoconservatism. See Otto's *The War on Science: Who's Waging It, What It Matters, What We Can Do about It* (Minneapolis: Milkweed Editions, 2016), p. 199.

7. Michel Foucault, "Truth and Power," in *The Foucault Reader*, ed. Paul Rabinow (New York: Pantheon Books, 1984), pp. 72–73.

8. Ibid., p. 70.

9. Paul Feyerabend, *Science in a Free Society* (London: New Left Books, 1978), p. 86.

10. Ibid., p. 87.

11. As this volume was going to press, Mouffe released a small volume arguing for a leftist populism that presents her and, to a certain extent, Laclau's advocacy of populism in a more accessible form. See Chantal Mouffe, *For a Left Populism* (New York: Verso, 2018).

12. Ernesto Laclau, *On Populist Reason* (London: Verso, 2005), p. 81.

13. Ibid.

14. Ibid., p. 154.

15. Jon Elster, "Hard and Soft Obscurantism in the Humanities and Social Sciences," *Diogenes* 58, no. 1–2 (February–May 2011): 159–70.

16. Consider for example the attack on intellectuals in the conservative historian Paul Johnson's *Intellectuals: From Marx and Tolstoy to Sartre and Chomsky* (New York: Harper & Row, 1988).

17. Russell Kirk, "Cultural Debris: A Mordant Last Word," in *The Portable Conservative Reader*, ed. Russell Kirk (New York: Penguin Books, 1996), p. 706.

18. See the debate from Buckley's *Firing Line* television show, "Firing Line Creation and Evolution Debate 1997," ChristopherHitchslap, October 19, 2011, YouTube video, 1:18:13, https://www.youtube.com/watch?v=9XZDTsQaxw8.

19. For this history, see George H. Nash, *The Conservative Intellectual Movement in America Since 1945* (Wilmington: ISI Books, 2006).

20. See, for example, Gabriella Coleman, *Hacker, Hoaxer, Whistleblower, Spy: The Many Faces of Anonymous* (New York: Verso Books, 2014). Angela Nagle has provided a similar criticism in her *Kill All Normies*.

21. For a discussion of right-wing internet culture, see Nagle, *Kill All Normies*. For an excellent, more general discussion of the effect of the internet on the notion of expertise, see Tom Nichols, *The Death of Expertise: The Campaign against Established Knowledge and Why It Matters* (Oxford: Oxford University Press, 2017), pp. 105–33.

22. For a discussion of "Red Pilling," see Abigail Brooks, "Popping the Red Pill: Inside a Digital Alternate Reality," CNNtech, November 10, 2017, http://money.cnn.com/2017/11/10/technology/culture/divided-we-code-red-pill/index.html.

23. For a sampling of Peterson's diatribes, see "Jordan Peterson on Women's Studies (from the Joe Rogan Experience #877)," PowerfulJRE, December 1, 2016, YouTube Video, 7:51, https://www.youtube.com/watch?v=88KJ5rgCNmk. A brilliant critique of Peterson and the irrational sources of his claims has been offered by Pankaj Mishra in "Jordan Peterson & Fascist Mysticism," *New York Review of Books*, March 19, 2018, http://www.nybooks.com/daily/2018/03/19/jordan-peterson-and-fascist-mysticism/.

24. For a discussion of the problem of "citizen journalism" and its right-wing practitioners, see Jesse Singal, "'Citizen Journalism' Is a Catastrophe Right Now, and It'll Only Get Worse," *New York Magazine*, October 19, 2016, http://nymag.com/selectall/2016/10/citizen-journalism-is-a-catastrophe-itll-only-get-worse.html.

25. For example, see "Citizen Journalists Are the Future of the Truth," Alex Jones, April 6, 2017, YouTube video, 5:09, https://www.youtube.com/watch?v=tvIDKUSFfXE [YouTube account has been taken down].

26. Steve Lohr, "It's True: False News Spreads Faster and Wider. And Humans Are to Blame," *New York Times*, March 8, 2018, https://www.nytimes.com/2018/03/08/technology/twitter-fake-news-research.html).

27. For a record of the debates and the efforts of creationists to make scientific arguments, see the sources collected in Robert T. Pennock and Michael Ruse, eds., *But Is It Science? The Philosophical Question in the Creation/Evolution Controversy*, updated ed. (Amherst, NY: Prometheus Books, 2009).

28. For more on Paltrow's claims, see Nichols, *Death of Expertise*, pp. 115–17.

29. For an excellent case in which Richard Dawkins took Chopra to task for his incorrect usage of scientific concepts, see "Richard Dawkins Interviews Deepak Chopra (Enemies of Reason Uncut Interviews)," Bernard Segura, May 4, 2013, YouTube Video, 22:12, https://www.youtube.com/watch?v=qsH1U7zSp7k.

30. For information on McCarthy's anti-vaccination campaigns, see Harry Collins, *Are We All Scientific Experts Now?* (Cambridge: Polity Press, 2014), pp. 111–12.

31. One of the best critiques of Žižek is Alan Johnson's "Slavoj Žižek's *Linksfaschismus*," in *Radical Intellectuals and the Subversion of Progressive Politics: The Betrayal of Politics*, ed. Gregory Smulewicz-Zucker and Michael J. Thompson (Basingstoke, UK: Palgrave Macmillan, 2015), pp. 99–120.

32. For more on the right-wing assault on science, see Chris Mooney, *The Republican War on Science* (New York: Basic Books, 2005) and Dave Levitan, *Not a Scientist: How Politicians Mistake, Misrepresent, and Utterly Mangle Science* (New York: W. W. Norton, 2017).

33. For a discussion, see Karen L. Cox, "The Whole Point of Confederate Monuments Is to Celebrate White Supremacy," *Washington Post*, August 16, 2017, https://www.washingtonpost.com/news/posteverything/wp/2017/08/16/the-whole-point-of-confederate-monuments-is-to-celebrate-white-supremacy/?utm_term=.304729db4cec.

34. Arlie Russell Hochschild, *Strangers in Their Own Land: Anger and Mourning on the American Right* (New York: New Press, 2016), p. 53.

35. See Rebecca Leber, "Making America Toxic Again," *Mother Jones*, March/April 2018, https://www.motherjones.com/politics/2018/02/scott-pruitt-profile-epa-trump/.

36. William Jennings Bryan, "Darwinism and the Schools," in *William Jennings Bryan: Selections*, ed. Ray Ginger (Indianapolis: Bobbs-Merrill, 1967), p. 237.

37. For video of Reagan's speech, see "James Robison: National Affairs Briefing (James Robison / LIFE Today)," Life Today TV, March 31, 2014, YouTube video, 25:57, https://www.youtube.com/watch?v=lH1e0xxRRbk.

38. Richard Feynman, "What Is Science?" in *The Pleasure of Finding Things Out: The Best Short Works of Richard P. Feynman*, ed. Jeffrey Robbins (Cambridge, MA: Perseus Books, 1999), p. 187.

39. Ibid., p. 186.

40. For a discussion of the relation between experts and the public, see Nichols, *Death of Expertise*, pp. 215–18. Nichols stresses the usefulness of experts. For a discussion of how experts in the natural sciences can serve as mediators between the public and policymakers, see Harry Collins and Robert Evans in their *Why Democracies Need Science* (Cambridge: Polity Press, 2017).

41. Thomas Frank, *Listen, Liberal: Or, What Ever Happened to the Party of the People?* (New York: Picador, 2017), p. 29.

42. Ibid., p. 39.

43. Richard Hofstadter, *Anti-Intellectualism in American Life* (New York: Vintage Books, 1963).

Chapter 11: Plato's Revenge: An Undemocratic Report from an Overheated Planet

*This essay was originally published in English in *Logos* 12, no. 2 (Spring 2013) and is used here with permission. Earlier, it had appeared in German as "*Platons Rache: Undemokratische Nachrichten von einem überhitzten Planeten*," in *Wissenschaft und Demokratie*, ed. Michael Hagner (Berlin: Suhrkamp, 2012), pp. 189–214.

1. The themes of many parts of this essay are developed at far greater length in my books *Science in a Democratic Society* (Amherst NY: Prometheus Books, 2011) and *The Seasons Alter: How to Save our Planet in Six Acts* (coauthored with Evelyn Fox Keller; New York: Norton/Liveright, 2017).

2. The professor was obviously thinking of the discussions in the *Republic*.

3. As I shall acknowledge later, even this is too optimistic. Virtually all climate scientists believe that, even if we act immediately, the global mean temperature will rise by 2–3°C by the end of the century.

4. For accessible presentations of their ideas, see James Hansen, *Storms of my Grandchildren: The Truth about the Coming Climate Catastrophe and Our Last Chance to Save Humanity* (New York: Bloomsbury, 2009); Stephen Henry Schneider, *Science as a Contact Sport: Inside the Battle to Save Earth's Climate* (Washington, DC: National Geographic, 2009); Michael E. Mann, *The Hockey Stick and the Climate Wars: Dispatches from the Front Lines* (New York: Columbia University Press, 2012).

5. Anthony Leiserowitz et al., *Climate Change in the American Mind: March 2018* (New Haven, CT: Yale Program on Climate Change Communication, 2018), p. 3, http://

climatecommunication.yale.edu/wp-content/uploads/2018/04/Climate-Change-American
-Mind-March-2018.pdf.

6. John Milton, *Areopagitica* (London: A. Millar, 1738); John Stuart Mill, *On Liberty* (London: John W. Parker and Son, West Strand, 1859).

7. Thomas S. Kuhn, *The Structure of Scientific Revolutions* (Chicago: University of Chicago Press, 1962), chap. XII.

8. John Dewey, *Democracy and Education* (New York: Free Press, 1944).

9. The ideal was originally introduced in my *Science, Truth, and Democracy* (New York: Oxford University Press, 2001). It is defended and developed much further in my *Science in a Democratic Society*.

10. The ideas of the following paragraphs are a *radical* compression of themes worked out far more extensively in my *Science in a Democratic Society* and in Kitcher and Keller *The Seasons Alter*. For the perspective on values, also see my book, *The Ethical Project* (Cambridge MA: Harvard University Press, 2011).

Chapter 12: Democracy and the Problem of Pseudoscience

1. See Michael Ruse, *Monad to Man: The Concept of Progress in Evolutionary Biology* (Cambridge, MA: Harvard University Press, 1996); Michael Ruse, *The Gaia Hypothesis: Science on a Pagan Planet* (Chicago: Chicago University Press, 2013); and, Michael Ruse, "Evolution: From Pseudoscience to Popular Science, from Popular Science to Professional Science," in *Philosophy of Pseudoscience: Reconsidering the Demarcation Problem*, ed. Mario Pigliucci and Maarten Boudry (Chicago: University of Chicago Press, 2013), pp. 225–45.

2. Ruse, *Monad to Man*.

3. Erasmus Darwin, *The Temple of Nature* (London: J. Johnson, 1803), pp. 26–28.

4. Erasmus Darwin, *Zoonomia; or, The Laws of Organic Life*, vol. 1 (London: J. Johnson, 1794), p. 509.

5. Michael Ruse, *The Darwinian Revolution: Science Red in Tooth and Claw* (Chicago: Chicago University Press, 1979).

6. These things are comparative. Had it been to do with football, Florida State could have gotten away with it, but not Harvard. See Ruse, *Gaia Hypothesis*.

7. Michael Ruse, ed., *But Is It Science? The Philosophical Question in the Creation/Evolution Controversy* (Amherst, NY: Prometheus Books, 1988).

8. See Michael Ruse, *The Philosophy of Biology* (London: Hutchinson, 1973); Michael Ruse, *The Darwinian Revolution: Why It Matters to Philosophers* (Cambridge: Cambridge University Press, 2018).

9. The definitive history is Ronald Numbers, *The Creationists: From Scientific Creationism to Intelligent Design* (Cambridge, MA: Harvard University Press, 2006). The book that got the modern movement kick-started was *Genesis Flood: The Biblical Record and Its Scientific Implications* (Philadelphia: Presbyterian and Reformed Publishing, 1961) by biblical scholar John C. Whitcomb Jr. and hydraulic engineer Henry M. Morris. My contribution back then was *Darwinism Defended: A Guide to the Evolution Controversies* (Reading, MA: Benjamin/Cummings, 1982).

10. Henry Morris et al., *Scientific Creationism* (San Diego, CA: Creation-Life, 1974).

11. See Michael Ruse, *Can a Darwinian Be a Christian? The Relationship between Science and Religion* (Cambridge, UK: Cambridge University Press, 2001).

12. Bill Clinton, the future president, was governor from 1978 to 1980. For the first and last time in his political career, he was complacent and got thrown out after one term. He came blasting back in 1982 and stayed governor until 1992, when he was elected president.

13. I don't think a religion is necessarily a pseudoscience or conversely that a pseudoscience is necessarily religious. I do think, as here, a pseudoscience can be produced for a religious end. Technically, what mattered in the court case was showing that religion was at work here. The First Amendment does not bar the teaching of pseudoscience in publicly funded schools. It bars the teaching of religion in such schools.

14. William Overton, "United States District Court Opinion: McLean versus Arkansas," in, Ruse, *But Is It Science?* pp. 307–31.

15. Phillip E. Johnson, *Darwin on Trial* (Washington, DC: Regnery Gateway, 1991).

16. Barbara Forrest and Paul R. Gross, *Creationism's Trojan Horse: The Wedge of Intelligent Design* (Oxford: Oxford University Press, 2004). No one thinks the intelligence is a grad student on Andromeda, playing games down here on earth as part of the dissertation project.

17. Michael Behe, *Darwin's Black Box: The Biochemical Challenge to Evolution* (New York: Free Press, 1996).

18. William Dembski, *The Design Inference: Eliminating Chance through Small Probabilities* (Cambridge, UK: Cambridge University Press, 1998); William Dembski, *Intelligent Design: The Bridge between Science and Theology* (Downer's Grove, IL: Intervarsity Press, 1999).

19. William Dembski and Michael Ruse, eds., *Debating Design: Darwin to DNA* (Cambridge, UK: Cambridge University Press, 2004).

20. Kenneth R. Miller, *Finding Darwin's God: A Scientist's Search for Common Ground between God and Evolution* (New York: Harper and Row, 1999).

21. Robert Pennock, *Tower of Babel: Scientific Evidence and the New Creationism* (Cambridge, MA: MIT Press, 1998).

22. Robert T. Pennock and Michael Ruse, eds., *But Is It Science? The Philosophical Question in the Creation/Evolution Controversy*, updated ed. (Amherst, NY: Prometheus Books, 2009). See especially, "Part III: Intelligent Design Creationism and the Kitzmiller Case."

23. Michael Ruse, *Darwinism as Religion: What Literature Tells Us about Evolution* (Oxford: Oxford University Press, 2017); Ruse, *Darwinian Revolution*.

24. Thomas Nagel, *Mind and Cosmos: Why the Materialist Neo-Darwinian Conception of Nature Is Almost Certainly False* (New York: Oxford University Press, 2012), p. 66.

25. Ibid.

26. James Lovelock, *Gaia: A New Look at Life on Earth* (Oxford: Oxford University Press, 1979).

27. See Ruse, *Gaia Hypothesis*.

28. John Postgate, "Gaia Gets Too Big for Her Boots," *New Scientist*, April 7, 1988, p. 60.

29. A question I am often asked is why I, a Brit working at an unfashionable university in Canada—we were, after all, the aggie college (still known as the "Cow College") with arts and sciences added on—was chosen by the ACLU to speak up for history and philosophy of science. Part of it was that I alone was willing to do the job. Most of my fellow philosophers thought that one should not take the ethereal refinement of the seminar into the vulgar spotlight of a federal court. After the trial, many people (fellow philosophers) criticized me for having spoken up as an expert witness. Apart from anything else, at the time it was unfashionable to agree with Karl Popper that one can find a criterion of demarcation between science and nonscience. And yet, this was precisely what I was doing. Part of the reason for being there was that the lawyers for the ACLU saw that I would be a good witness. Years of teaching first-year undergraduates—something I still do with great joy and a sense of privilege—had honed my skills at speaking clearly and forcefully. Most importantly, I fully realized that a good joke is worth a thousand arguments. At some point, the assistant district attorney was badgering me about my religious beliefs. Finally, I blurted out, "Mr. Williams. Can't you see that I am not an expert witness on my religious beliefs?" Everyone burst into laughter and then, when Williams returned to the attack,

the judge said, "Mr. Williams. Can't you see that he is not going to give you what you want? Just move on."

30. Actually, I have touched on environmental issues in some of my recent writings, notably my book *The Gaia Hypothesis*, and a recent book on science and religion coauthored by the Pulitzer Prize winner, Edward J. Larson, *On Faith and Science* (New Haven, CT: Yale University Press, 2017).

31. Philip Kitcher, *Abusing Science: The Case against Creationism* (Cambridge, MA: MIT Press, 1982).

32. Michael Ruse *Darwinism and Its Discontents* (Cambridge, UK: Cambridge University Press, 2006).

33. Michael Ruse, *The Evolution-Creation Struggle* (Cambridge, MA: Harvard University Press, 2005).

34. Michael Ruse, *Darwinism as Religion: What Literature Tells Us about Evolution* (Oxford: Oxford University Press, 2017).

35. Michael Ruse, *The Problem of War: Darwinism, Christianity, and Their Battle to Understand Human Conflict* (Oxford: Oxford University Press, 2018).

36. Michael Ruse, *A Meaning to Life* (Oxford: Oxford University Press, 2019).

37. Michael Ruse, *On Purpose* (Princeton, NJ: Princeton University Press, 2017).

38. If someone wants to make a religion out of Darwinian progress, as I think the eminent evolutionist Edward O. Wilson—*On Human Nature* (Cambridge, MA: Harvard University Press, 1978); Edward O. Wilson, *The Creation: A Meeting of Science and Religion* (New York: Norton, 2006); Edward O. Wilson, *The Meaning of Human Existence* (New York: Liveright, 2014)—attempts, then I don't want to go down that path, but I am not about to stop others. My worry is that, as happens too frequently, people start to think that this is a valid branch of science. It isn't.

39. Jean-Paul Sartre, *Existentialism and Humanism*, trans. Philip Mairet (Brooklyn, NY: Haskell House, 1977), p. 56.

40. Ibid., pp. 27–28.

41. Michael Ruse, *The Philosophy of Human Evolution* (Cambridge, UK: Cambridge University Press, 2012).

42. Michael Ruse, *Atheism: What Everyone Needs to Know* (Oxford: Oxford University Press, 2015).